Lectures in Abstract Algebra

Basic theory on Groups and Rings with Gröbner bases

Dong-il Lee

Kyowoo Publishing Company

Dong-il Lee

Education :
- Ph.D.(Integrated), Seoul National University, February 2008
- B.S., Seoul National University, August 2002

Career :
- Assistant Professor, Department of Mathematics, Seoul Women's University (September 2009 -)
- Research Fellow, School of Computational Sciences, Korea Institute for Advanced Study (March 2008-August 2009)

Kyowoo Publishing Company
8, Yangnyeongsi-ro, Dongdaemun-gu,
Seoul 130-861 Korea

Phone: 82-2-925-2861
F a x: 82-2-925-2860
e-mail: kyowoo@kyowoo.co.kr
http://www.kyowoo.co.kr

ISBN 979-11-251-0056-0 (93410)
Printed in the Republic of korea.

Preface

There is no doubt that we have to keep learning for a lifetime. One of the hardest question to answer is: Why do we study mathematics? ...[1]

First of all, let us answer the question, *What is Mathematics?* What does the word, mathematics, originate from? That stems from 'mathesis', meaning 'mental culture (情神 修養)' or 'learning'.[2] Mathematics has a long history of intellectual exploration of human beings. So, studying math is of a great help to understanding human nature and eventually to making our life happy. From a more practical point of view, it is a foundation of modern civilization and a core of every scientific activity.

What is *(Abstract) Algebra*? Oxford English Dictionary says that (Abstract) Algebra is a branch of mathematics concerning the study of the rules of operations. e.g.) the addition in the set of natural numbers, the multiplication, ... We study properties of binary operations. Trying to solve quintic equations leads to the study of groups and fields, followed by Galois theory in 19C.

The Arabic word *al-jabr* presumably meant something like "restoration" or "completion" and seems to refer to the transposition of subtracted terms to the other side of an equation.

We want to study the algebraic structures. An algebraic structure is a set with operations.

e.g.) vector space, group, ring, field, module, ...

Why to study *Abstract*? Easy! Essential(=precise)! General(=powerful)! Leading to more *concrete* understanding.

As a concrete example, we consider a regular triangle R in the plane, and find all transformations which make R invariant. We have a correspondence between an object in a space and the set of transformations. Note that a given object must

[1] Some people say that *curiosity* is an essential motivation for studying mathematics.

[2] `http://www.math.snu.ac.kr/~hongjong/c4ns/Math.htm`

be invariant under a transformation, so we say that it has a *symmetry* and that the set of transformations which leave an object invariant has a "good" structure. Felix Klein, a German mathematician in 19C, said that Geometry is the study of symmetry. Modern Algebra is the study of a good structure, such as the collection of transformations which make a regular polygon invariant.

A main focus of classical Algebra was solving polynomial equations. On the other hand, as the study of algebraic structures becomes more important since 19th century, modern Algebra may be defined to be studying the beautiful structure of operations.
- It's very interesting that Diophantus[3] once described the solution of $4x + 6 = 2$ as an absurd number. For hundreds of years even until 18C, mathematicians refused to accept the complex numbers(numbers which involve $\sqrt{-1}$) because they couldn't find concrete examples or applications, although they are now basic tools of physics. Throughout the history, the scope of our understanding has been extended.
- There are many great mathematicians (Fermat, Euler, Lagrange, Gauss, Abel, Galois, Kronecker, Noether, ...)[4] who set a foundation of Algebra from 17th century until the early 20C.
- One of the greatest achievements in the early 19C is Abel's proof for the insolvability of a general quintic equation in an algebraic way.
- Linear and Abstract Algebra is a fundamental language in mathematics and science, and becomes basics in almost all areas of mathematics, including applied mathematics like coding theory, cryptography, computational algebra, etc.

This note grew out of lectures for the course at the department of mathematics, Seoul Women's University. The lecture is for undergraduate students majoring in mathematics, who first encounter an abstract theory of algebraic structures.

The contents of this note are mostly basic and fundamental ones. The items attached with a star * are advanced materials, so they may be skipped at the first reading.

---

[3]Diophantus(210?–290?), an Alexandrian Greek mathematician and the author of a series of books called Arithmetica.

[4]For their portraits, refer to the front cover in [이II].

Up to Chapter 3 in this note is a lecture note of the 1st semester, whose title is *Basic group theory.* Group theory is a language, indispensable for studying modern mathematics. The amount to be covered in the first semester is most of the Part I–III, and some of the Part VII in the textbook [F]:

symmetric groups, homomorphisms, quotient groups, simple groups, ...

The title of the second course is *Basic theory on Rings and Fields.* We deal with Part IV–VI and IX in the textbook [F]:

ideal theory, polynomials, field extensions, finite fields, ...

One of the main objects to be studied is the ring of *polynomials.* A computation involving polynomials can be performed by a software. For example, the computer algebra system SINGULAR [5] provides the command `factorize(f);` for factorizing a polynomial as follows:

```
> ring R=0,(x,y),dp;
> poly f=x2+5xy+6y2;
> factorize(f);
```

We included several interesting applications using basic theory:

error-correcting codes, (extended) Euclidean algorithm, quaternions,

geometric constructions, algebraic equations, ...

For a taste of the beautiful connection between fields and groups, a part of

Galois theory

is briefly written down usually in the form of Exercises. Also the students full of curiosity are invited to study the theory of Computational Algebra for

multivariate polynomials,

which is described in the final chapter as elementary as possible to show so many powerful applications.

This note contains regular exam problems in the course of Abstract Algebra for 5 years as references for undergraduate students.

[5] Available at `http://www.singular.uni-kl.de/` . In 2006, Decker and Lossen published an introductory textbook [DL] based on SINGULAR.

I am grateful for the friendship and support I've had from my colleagues since my first semester in 2009 at Seoul Women's University. Warm thanks should be given to the students who attended my classes. Finally, I close this prologue with my sincere heart toward J...

Dong-il Lee[6]
Seoul, Korea
August 2014

[6] This work was supported by a research grant from Seoul Women's University(2013).

Contents

Preliminaries

We use common notations throughout this book:

$\mathbb{Z}$,	the set of integers,
$\mathbb{Z}_{\geq 0}$,	the set of nonnegative integers,
$\mathbb{Z}_{+}$,	the set of positive integers,
$\mathbb{Q}$,	the set of rational numbers,
$\mathbb{R}$,	the set of real numbers,
$\mathbb{C}$,	the set of complex numbers.

In this chapter, we introduce some basic notions which will be used in the subsequent chapters.

1. Partitions and equivalence relations

Partitions and equivalence relations are defined in a different way, however, we'll see that they are the same notions.

DEFINITION 1.1. Sets A and B are disjoint if $A \cap B = \emptyset$. A *partition* of a set S is a collection of nonempty subsets A_i's ($i \in I$) such that A_i's are disjoint each other and $\bigcup_{i \in I} A_i = S$. Here, the subsets A_i's are called the cells of the partition of S, and we denote the cell containing $x \in S$ by $[x]$ or $\overline{x}$.

Later, we'll see that it is important that we divide a set into its disjoint subsets.

EXAMPLE 1.2 (A partition of $\mathbb{Z}$ into n cells). Let $\mathbb{Z}$ be the set of integers and n be a positive integer. Define $\overline{r} = \{nk + r \mid k \in \mathbb{Z}\}$ (say $r = 0, 1, \ldots, n-1$). Then these cells are called the *residue classes modulo* n in $\mathbb{Z}$, which form a partition of $\mathbb{Z}$. Note that if $n = 3$ then

$$\overline{0} = \{3k \mid k \in \mathbb{Z}\} = \{\ldots, -3, 0, 3, 6, 9, \ldots\} = \overline{3} = \overline{6}, \text{ and}$$

$$\overline{1} = \{\ldots, -2, 1, 4, 7, 10, \ldots\} = \overline{4} = \overline{-2}, \quad \overline{2} = \overline{-1},$$

Thus, $\{\overline{0}, \overline{1}, \overline{2}\}$ is a partition of $\mathbb{Z}$.

DEFINITION 1.3. (a) A relation R on a set S is a subset of $S \times S$. We say that $x \in S$ is related to $y \in S$, usually denoted by $x \sim y$, if $(x, y) \in R$.

(b) A relation R on a set S is called an *equivalence relation* on S if it satisfies the following conditions:

(i) (reflexive) $x \sim x$ for each $x \in S$;

(ii) (symmetric) $x \sim y$ implies $y \sim x$ for each $x, y \in S$;

(iii) (transitive) $x \sim y$ and $y \sim z$ imply $x \sim z$ for each $x, y, z \in S$.

EXAMPLE 1.4. Let $R = \{(x, y) \in \mathbb{Z} \times \mathbb{Z} | \ x \geq y\}$. Then R is a relation on $\mathbb{Z}$, but it is not an equivalence relation.

REMARK 1.5. (c.f.) A relation $\leq$ on a set S which are reflexive, antisymmetric[7] and transitive is called a *partial order* on S. We call a set S with a partial order $\leq$ to be a *partially ordered set* (abbreviated as a *poset*). A partial order $\leq$ on S is called a *total order*(also a linear order or a simple order) if every two elements are comparable.

EXAMPLE 1.6 (Congruence in $\mathbb{Z}$). Let n be a positive integer. An integer a is *congruent* to an integer b modulo n, denoted by

$$a \equiv b \pmod{n}$$

if a and b are in the same $\overline{r} = \{nk + r | \ k \in \mathbb{Z}\}$ for some r, i.e., $a - b$ is divisible by n. Then the congruence $\equiv$ is an equivalence relation.

REMARK 1.7. (1) An equivalence relation is an abstraction of the equality $=$.

(2) Each partition of a set S yields an equivalence relation $\sim$ on S by defining $x \sim y$ if and only if $\overline{x} = \overline{y}$, that is, x and y are in the same cell of the partition.

Now we prove that any equivalence relation induces a partition by the equivalence classes.

Theorem 1.8. *Let $\sim$ be an equivalence relation on a set S. We define the* equivalence class *containing $a \in S$,*

$$[a] = \overline{a} := \{x \in S | \ x \sim a\}.$$

[7]A relation $\sim$ on S is called *antisymmetric* if, for each $x, y \in S$, $x \sim y$ and $y \sim x$ implies $x = y$.

The element a is called a representative *of the class $\overline{a}$. Then $\{\overline{a}|\ a \in S\}$ is a partition of S. That is, every element of S is in exactly one equivalence class.*

Proof. Let $a \in S$. First, $a \in \overline{a}$ since $a \sim a$. We show that if $a \in \overline{b}$ then $\overline{a} = \overline{b}$. Assume $a \in \overline{b}$, i.e., $a \sim b$. Then we show $\overline{a} \subseteq \overline{b}$ and $\overline{a} \supseteq \overline{b}$. Let $x \in \overline{a}$, i.e., $x \sim a$. Together with $a \sim b$, we have $x \sim b$, i.e., $x \in \overline{b}$. Conversely, let $y \in \overline{b}$, i.e., $y \sim b$. Together with $b \sim a$ from $a \sim b$, we have $y \sim a$, i.e., $a \in \overline{a}$. So we've proved that $\overline{a} = \overline{b}$. □

Example 1.9. Let $S = \{\text{members in this Algebra class}\}$. For $x, y \in S$, $x \sim y$ is defined as: x and y have the same gender. Then $S = \{\text{males}\} \coprod \{\text{females}\}$.

Exercise 1.10. (1) Which of the following subsets of $\mathbb{R} \times \mathbb{R}$ are equivalence relations on $\mathbb{R}$?

1) $\{(x, y)|\ x - y \text{ is an even integer}\}$;
2) $\{(x, y)|\ x - y \text{ is rational}\}$;
3) $\{(x, y)|\ x + y \text{ is rational}\}$;
4) $\{(x, y)|\ x - y \leq 0\}$.

(2) Do any of the following define equivalence relations on the set of all nonzero complex numbers?

1) z is related to w if zw is real;
2) z is related to w if z/w is real;
3) z is related to w if zw is an integer.

2. Binary operations

The usual addition and multiplication are typical examples of binary operations.

Definition 2.1. A *binary operation* $*$ on a set S is a mapping $* : S \times S \to S$. Usually, we denote $*(a, b)$ by $a * b$. If a set S is finite then a binary operation $*$ on S can be represented by a finite table.

Example 2.2. Let $\mathbb{N} = \{0, 1, 2, \ldots\}$ be the set of nonnegative integers. The usual addition $+$ is a mapping $+ : \mathbb{N} \times \mathbb{N} \to \mathbb{N}$. Also, $+$ is defined on the set $2\mathbb{N}$ of even nonnegative integers.

DEFINITION 2.3. Let $(S, *)$ and $(S', *')$ be sets with binary operations. An *isomorphism* ϕ of S with S' is a 1-1 function mapping S onto S' such that

$$\phi(a * b) = \phi(a) *' \phi(b) \quad \text{for all } a, b, \in S.$$

In this case, we say that $(S, *)$ is *isomorphic* to $(S', *')$ and denote by $(S, *) \cong (S', *')$.

REMARK 2.4. Note that $\cong$ is an equivalence relation.

Our question is:

How to determine whether two sets with binary operations are isomorphic or not? It's not easy to check, in general.

EXAMPLE 2.5. (1) We see easily that $(\mathbb{N}, +) \cong (2\mathbb{N}, +)$ through an isomorphism ϕ given by $\phi(x) = 2x$ for $x \in \mathbb{N}$.

In general, let $\mathbb{Z}$ be the set of integers and n be a positive integer. Define a map ϕ from $n\mathbb{Z} = \{nk |\ k \in \mathbb{Z}\}$ to $\mathbb{Z}$ by $\phi(nk) = k$. Then ϕ is an isomorphism of $(n\mathbb{Z}, +)$ with $(\mathbb{Z}, +)$ since it is 1-1, onto and $\phi(na+nb) = \phi\left(n(a+b)\right) = a+b = \phi(na)+\phi(nb)$.

(2) Let

$$U_n = \{z \in \mathbb{C} |\ z^n = 1\} = \{\zeta_n^i |\ i = 0, 1, 2, \ldots, n-1\}$$

where $\zeta_n = e^{\frac{2\pi i}{n}} = \cos\frac{2\pi}{n} + i \sin\frac{2\pi}{n}$. The usual multiplication is defined on U_n. On the other hand, consider

$$\mathbb{Z}_n = \{\overline{0}, \overline{1}, \overline{2}, \ldots, \overline{n-1}\}$$

where $\overline{r} = \{nk + r |\ k \in \mathbb{Z}\}$ with $\overline{r} + \overline{s} = \overline{r+s}$. Then U_n with the multiplication is isomorphic to $\mathbb{Z}_n$ with the addition. Explicitly, $\zeta_n \in U_n$ corresponds to $\overline{1} \in \mathbb{Z}_n$. We'll see that these are *cyclic* groups of order n, in the next chapter.

(3) Let $H = \left\{ \begin{pmatrix} a & b \\ -b & a \end{pmatrix} |\ a, b \in \mathbb{R} \right\}$. Then we can check that H is closed under the matrix addition and multiplication. Furthermore, if we define $\phi : \mathbb{C} \to H \subset M_{2\times 2}(\mathbb{R})$ to be

$$\phi(a + bi) = \begin{pmatrix} a & b \\ -b & a \end{pmatrix},$$

then we know that $(\mathbb{C}, +) \cong (H, +)$ and $(\mathbb{C}, \cdot) \cong (H, \cdot)$ since ϕ is 1-1, onto and

$$\phi\left((a + bi) + (a' + b'i)\right) = \phi(a + bi) + \phi(a' + b'i),$$

$$\phi\left((a + bi) \cdot (a' + b'i)\right) = \phi(a + bi) \cdot \phi(a' + b'i).$$

EXERCISE 2.6. We define the addition on $\mathbb{Z} \times \mathbb{Z}$ by $(x, y) + (x', y') = (x + x', y + y')$. Is $\mathbb{Z} \times \mathbb{Z}$ isomorphic to $\mathbb{Z}$?

We introduce two rules of binary operations.

DEFINITION 2.7. Let S be a set equipped with a binary operation $*$.

(a) A binary operation $*$ on S is *associative* if $(a*b)*c = a*(b*c)$ for all $a, b, c \in S$.

(b) A binary operation $*$ on S is *commutative* if $a*b = b*a$ for all $a, b \in S$.

EXAMPLE 2.8. (1) The usual addition $+ : \mathbb{N}\times\mathbb{N} \to \mathbb{N}$ is commutative and associative.

(2) Let f, g, h be functions mapping a set S into S. Then $f \circ (g \circ h) = (f \circ g) \circ h$, that is, the composition of functions is associative.

(3) Let $M_{2\times 2}(\mathbb{R})$ be the set of all 2×2 matrices with real entries. Since the multiplication of matrices corresponds to the composition of linear functions, the matrix multiplication is associative. Is the matrix multiplication on $M_{2\times 2}(\mathbb{R})$ commutative?

(4)* The multiplication for quaternions is not commutative. A benefit of inventing and using quaternions will be presented in Ch.4 §6.

(5)* The multiplication for octonions is not associative.[8]

[8]For details, see [Bz].

CHAPTER 1

Basic notions of groups

Several origins of the notion of groups are
1) the transformation groups coming from geometry in the 19th century, [Ev, §30–§31]
2) classical number theory due to Euler[1] and Gauss[2], and
3) solving algebraic equations which leads Lagrange[3] to the study of permutations of the roots of an equation. [F, p.38] [4]

Let us give more details on 1).

EXAMPLE. Consider a regular triangle T in $\mathbb{R}^2$. What are all transformations on $\mathbb{R}^2$ which make T invariant?

Some focuses of 19C geometry are searching invariants under geometric transformations, such as the symmetry of polytopes, and studying transformations themselves. A set G consisting of self-bijections on a set S is called a *transformation group* on S if it satisfies:

i) $\mathrm{id} \in G$;

ii) $g \in G \implies g^{-1} \in G$;

iii) $g_1, g_2 \in G \implies g_1 \circ g_2 \in G$.

---

[1]Leonhard Paul Euler(1707–1783), a pioneering Swiss mathematician.

[2]Carl Friedrich Gauss(1777–1855), a German mathematician and scientist.

[3]Joseph Louis Lagrange(1736–1813), a French mathematician and astronomer.

[4]Root formulas for cubic and quartic equations were discovered by Italian mathematicians in 16C. For quintic equations, in 19C, Abel(1802–1829, a Norwegian mathematician) proved that they are not solvable by radicals, and Galois(1811–1832, a French mathematician) developed more systematic theory on the correspondence between fixed fields and automorphism groups. [김김, HP5]

Geometry can be defined to be the study of properties invariant under a transformation group.[5] That is, geometry is to study a kind of "symmetry".[6] We observe that a collection of transformations forms a nice structure with an operation. This is the notion of a group. Group theory is studying an abstraction of symmetry, and *modern algebra is to study abstract algebraic structures.*

One of the Hilbert's 18th problems[7] is the classification of crystals in the n-dimensional space. A *crystal* is a geometric structure which is periodic [8] and discrete. Note that

$$\{\phi, \text{isometry} \mid \phi(S) = S \text{ for } S \subseteq \mathbb{R}^2\} \leq \{\text{isometries of } \mathbb{R}^2\} \leq \{\text{permutations of } \mathbb{R}^2\}.$$

The plane isometries are translations, rotations, reflections, and glide reflections [F, §12]. The *strip patterns* or the frieze groups in the plane are divided into 7 types.[9]

1. Groups

We introduce the first object of algebraic structures.

DEFINITION 1.1. A *group*[10] $(G, *)$ is a set G with a binary operation $*$, satisfying that

0) G is closed under $*$;

1) $*$ is associative;

[5]Due to Felix Klein(1849–1925). Euclidean geometry is the study of invariants under Euclidean motions. Klein gave a famous inaugural lecture at Erlangen University in 1872.

[6]Harmony of proportions, or a feeling of beauty. Geometrically, we say that an object has a symmetry if it is invariant under a transformation. For a general account, recommend to read four lectures [We] by Hermann Weyl(1885–1955).

[7]David Hilbert(1862–1943), one of the most influential mathematicians in the history of mathematics, presented 23 problems in the International Congress of Mathematicians held in Paris 1900. The problems made a strong drive on the development of 20C mathematics.

[8]After the discovery of quasicrystals(which are ordered but not periodic) by D. Shechtman(2011 Nobel prize in Chemistry) in 1982, the notion of crystals is redefined in the early 1990's. For details on the new definition, refer to `http://reference.iucr.org/dictionary/Crystal`. A typical example of an aperiodic tiling is the Penrose tiling. One finds this example in the book [Gd].

[9]걷기, 외발뛰기, 옆걸음, 두발뛰기, 돌며외발뛰기, 돌며옆걸음, 돌며두발뛰기. For the explicit frieze groups, refer to [김김, p.256–257]. The wallpaper groups or the plane crystallographic groups are divided into 17 types [G, Ch.30]. There are 230 three-dimensional crystallographic groups. The four-dimensional space has 4783 ones.

[10]We call it 군(群) in Korean.

2) G has an identity element, that is,
there is an element $e \in G$ such that $a * e = a = e * a$ for all $a \in G$;

3) For each $a \in G$, G has an inverse element of a, that is,
for each $a \in G$, there is $b \in G$ such that $a * b = e = b * a$.

A group is called *abelian*[11]if its binary operation is commutative.

DEFINITION 1.2. We denote the number of elements in a group G by $|G|$, called the *order* of G. A group G is a finite group if $|G| < \infty$, and G is an infinite group if $|G| = \infty$.

EXAMPLE 1.3. (1) The set $\mathbb{Z}$ of integers with the usual addition $+$ is an infinite abelian group.

(2) The set $\mathbb{Q}^{\times}$ of nonzero rational numbers with the usual multiplication is an infinite abelian group.

(3) The set

$$\mathbb{Z}_n = \{\overline{0}, \overline{1}, \overline{2}, \ldots, \overline{n-1}\}$$

with the modular addition is a finite abelian group. Note that $(\mathbb{Z}_n, +)$ is isomorphic to the group $(U_n, \times)$ of nth roots of 1.

(4) Consider the modular multiplication in $\mathbb{Z}_n$. For instance, $\overline{2} \cdot \overline{3} = \overline{6} = \overline{1} \in \mathbb{Z}_5$. The set $\{\overline{1}, \overline{2}, \overline{3}, \overline{4}\} \subset \mathbb{Z}_5$ with multiplication is a group, and $\{\overline{1}, \overline{3}, \overline{5}, \overline{7}\} \subset \mathbb{Z}_8$ with multiplication is a group. In general, the set $\mathbb{Z}_n^{\times} = \{\overline{k} | \gcd(k, n) = 1\}$ is a group under multiplication, called the *unit group* of $\mathbb{Z}_n$. Note that $(\mathbb{Z}_4, +) \cong (\mathbb{Z}_5^{\times}, \times)$, and also $|\mathbb{Z}_8^{\times}| = 4$, but we'll see that

$$(\mathbb{Z}_4, +) \ncong (\mathbb{Z}_8^{\times}, \times).$$

(5) Two groups $(\mathbb{R}, +)$ and $(\mathbb{R}_+, \cdot)$ are isomorphic since the map $\exp : \mathbb{R} \to \mathbb{R}_+$ or $\log : \mathbb{R}_+ \to \mathbb{R}$ is an isomorphism between $(\mathbb{R}, +)$ and $(\mathbb{R}_+, \cdot)$.

(6) The set of translations in $\mathbb{R}^2$, that is,

$$G = \{T_{h,k} : \mathbb{R}^2 \to \mathbb{R}^2, (x, y) \mapsto (x + h, y + k) | \ h, k \in \mathbb{R}\}$$

with the composition $\circ$ becomes an infinite group.

(7) The set of rotations in $\mathbb{R}^2$ by $\frac{k\pi}{3}$ $(k = 0, 1, 2, 3, 4, 5)$ with $\circ$ is isomorphic to $(\mathbb{Z}_6, +)$ as a group.

[11]This term was named in honor of the Norwegian mathematician Niels Henrik Abel(1802–1829) [F, p.39].

(8) Let X be a set. If U and V are subsets of X, define $U \setminus V = \{x \in U | \; x \notin V\}$. The *Boolean group* $B(X)$ [12] is the family of all subsets of X equipped with addition given by symmetric difference $A + B$, where

$$A + B = (A \setminus B) \cup (B \setminus A).$$

(9) We call a self-bijective map to be a *permutation*. Consider

$$S_n = \{\sigma | \; \sigma \text{ is a permutation of } \{1, 2, \ldots, n\}\}$$

with the composition[13], which is called the nth *symmetric group*. Note that S_n is not an abelian group if $n \geq 3$. In fact,

$$S_3 = \{\text{id}, (1,2), (2,3), (1,3), (1,2,3), (1,3,2)\}$$

is a nonabelian group because $(1,2) \circ (2,3) = (1,2,3)$ but $(2,3) \circ (1,2) = (1,3,2)$, and we will see that S_3 is the smallest nonabelian group.

EXAMPLE 1.4. Let $\mathbb{F}$ be a field[14]. Then we define

$$GL_n(\mathbb{F}) := \{A \in M_{n \times n}(\mathbb{F}) \mid A \text{ is invertible } \},$$

which is a group under matrix multiplication[15], called the *general linear group* of degree n over $\mathbb{F}$.

EXERCISE 1.5. Let

$$G = \left\{ x, \frac{1}{x}, 1 - x, \frac{1}{1-x}, \frac{x-1}{x}, \frac{x}{x-1} \right\}$$

where $a * b$ is defined to be the replacement of x in a by b. For instance,

$$(1-x) * \frac{x}{x-1} = 1 - \frac{x}{x-1} = \frac{1}{1-x}.$$

Then show that $(G, *)$ is isomorphic to $(S_3, \circ)$ as a group.

---

[12]This was named after the English logician G. Boole(1815–1864).

[13]A permutation is represented by "사다리표". We compose two permutations by attaching two ladders, and the inverse permutation is obtained by reversing the ladder. [김김, HP5]

[14]Roughly speaking, a *field* is an algebraic structure which is closed under addition, subtraction, multiplication, and division. The name represents 운동장 or 마당, 즉 자유로운 공간, but we call it 체(體) meaning 몸, which originates from Körper in German. Usually, we take $\mathbb{F} = \mathbb{Q}, \mathbb{R}$ or $\mathbb{C}$. The finite field $\mathbb{F}_q$ with order $q = p^n$ for any prime p, plays an important role in the process of computers we are using everyday.

[15]For those who are not familiar with the matrix operations, read pp. 477–481 Appendix: Matrix Algebra in the book [F].

Proposition 1.6. *Let $(G, *)$ be a group. Then*

(a) *The left(right) cancellation law holds. Thus, in the operation table of G, there appears one and only one element at each row and column.*

(b) *An identity element e is unique.*

(c) *For each $g \in G$, an inverse g^{-1} of g is unique. As a corollary, for all $a, b \in G$,*

$$(a * b)^{-1} = b^{-1} * a^{-1}.$$

Proof. (a) For each $a, b, c \in G$, $a * b = a * c$ implies $b = c$ by multiplying an inverse of a from the left. Likewise, $b * a = c * a$ implies $b = c$. So if we take contrapositions, we know that each row and column in the operation table has mutually distinct elements.

(b) Suppose that there are identities e and e'. Then $e = e' * e = e'$.

(c) For $a \in G$, suppose that there are inverses a' and a''. Then $a * a' = a * a'' = e$ implies $a' = a''$ by the cancellation law. The corollary follows from $(a * b) * (b^{-1} * a^{-1}) = a * (b * b^{-1}) * a^{-1} = (a * e) * a^{-1} = a * a^{-1} = e$. □

REMARK 1.7. * (1) The definition 1.1 of a group is enough if the conditions 2) and 3) are just one-sided.

(2) We say that $(G, *)$ is a *semigroup* if it satisfies only 0) and 1), and a *monoid* it satisfies 0), 1) and 2) in Definition 1.1. The set $(\mathbb{N}, +)$ is a monoid. Let $[X]$ be the set of of all commutative monomials on a set of alphabets X, then $[X]$ is a monoid with the concatenation. For instance, if $X = \{x, y\}$ then $[X] = \{1, x, y, x^2, xy, y^2, x^3, x^2y, xy^2, y^3, x^4, \ldots\}$.

EXAMPLE 1.8. For the groups of order ≤ 3, there is only one group structure. How about those of order 4?

$*$	e	a	b	c
e	e	a	b	c
a	a	b	c	e
b	b	c	e	a
c	c	e	a	b

,

$*$	e	a	b	c
e	e	a	b	c
a	a	e	c	b
b	b	c	e	a
c	c	b	a	e

.

Note that the left table is isomorphic to that of $\mathbb{Z}_4 = \{\bar{0}, \bar{1}, \bar{2}, \bar{3}\}$ under the modular addition. The right one is called the *Klein 4-group*, sometimes denoted by V. [16]

[16] This notation comes from the German word Vier, meaning four.

Note that the right table is isomorphic to that of

$$\mathbb{Z}_2 \times \mathbb{Z}_2 = \{(\overline{0},\overline{0}),(\overline{1},\overline{0}),(\overline{0},\overline{1}),(\overline{1},\overline{1})\}$$

under the componentwise modular addition.

REMARK 1.9. (1) Later, we'll see that every group of order p (p, a prime) is isomorphic, that is, there exists only one group of order p up to isomorphism. So

$$(\mathbb{Z}_5, +)$$

is the only group of order 5 up to isomorphism.

(2) Also, it will be clear that we have two non-isomorphic groups of order 6. They are

$$(\mathbb{Z}_6, +) \not\cong (S_3, \circ).$$

If we have enough knowledge for the group theory, we could determine every possible structures at least for groups of small order as listed in [DF, §5.3] or [Hm, Appendix].

EXERCISE 1.10. Let G be a group. Show that if $a * a = e$ for all $a \in G$ (i.e., each element is an *involution*), then G is abelian.

Solution. Since $(a * b)^2 = e = (a * a) * (b * b)$ for all $a, b \in G$, the cancellation law shows that $a * b = b * a$, i.e., G is abelian. □

EXERCISE 1.11. Show that if G is a finite group with identity e and with an even number of elements, then there is $a \neq e$ in G such that $a * a = e$.

Solution. Suppose that there is no $a \in G$ such that $a \neq e$ and $a * a = e$. Then, for any non-identity $x \in G$, there exists $y = x^{-1} \neq x$ in G such that $x * y = y * x = e$. So the number of non-identity elements in G is even, thus $|G|$ is odd, which is a contradiction. Hence, there must be a non-identity $a \in G$ such that $a * a = e$. □

EXERCISE 1.12. Let G be a group and let g be a fixed element of G. Show that the map i_g defined by

$$i_g(x) = gxg^{-1} \quad \text{for } x \in G$$

is an isomorphism of G with itself.

2. Subgroups

It's very natural to introduce the notion of a subobject from that of an object.

DEFINITION 2.1. Let $(G, *)$ be a group. If a subset H of G is a group under the same operation $*$, then we say that H is a *subgroup* of G and denote by $H \leq G$. We call $H \lneq G$ a proper subgroup of G and the subgroup $\{e\} \leq G$ is called the *trivial* subgroup.

How to determine whether a subset H of a group $(G, *)$ is a subgroup or not?

Proposition 2.2. *A subset H is a subgroup of $(G, *)$ if and only if i) H is closed under $*$, ii) $e_G \in H$, and iii) $a^{-1} \in H$ for all $a \in H$. As a corollary, we know that $H \leq G$ if and only $a * b^{-1} \in H$ for all $a, b \in H$.*

EXAMPLE 2.3. (1) $\mathbb{Z} \leq \mathbb{Q} \leq \mathbb{R} \leq \mathbb{C}$ under the usual addition $+$.

(2) Let $F = \{f \mid f : [0,1] \to \mathbb{R}\}$, $C = \{f \mid f : [0,1] \to \mathbb{R}, \text{ continuous}\}$, and $D = \{f \mid f : [0,1] \to \mathbb{R}, \text{ differentiable}\}$. Then $D \leq C \leq F$.

(3) For any positive integer n, the set

$$U_n = \{\zeta^k \mid \zeta = \cos\frac{2\pi}{n} + i\sin\frac{2\pi}{n}, k = 0, 1, 2, \ldots, n-1\}$$

of all nth roots of unity forms a group under the usual multiplication. The elements of U_n can be represented by equally spaced points on a circle about the origin. How many different subgroups do you find when $n = 6$?

EXERCISE 2.4. (1) Consider the additive group $\mathbb{Z}_6 = \{\overline{0}, \overline{1}, \overline{2}, \overline{3}, \overline{4}, \overline{5}\}$ of 6 elements. Find the subgroups of $\mathbb{Z}_6$ and give the subgroup diagram for these subgroups of $\mathbb{Z}_6$.

(2) What are the subgroups of $(S_3, \circ)$?

(3)* Consider the 4th dihedral group $D_{2\cdot 4}$, the group of symmetries of a square. Could you find the subgroup diagram of $D_{2\cdot 4}$?

EXAMPLE 2.5. (1) Consider a subset

$$SL_n(\mathbb{R}) := \{A \in M_{n\times n}(\mathbb{R}) \mid \det A = 1\}$$

of $GL_n(\mathbb{R})$. Then $SL_n(\mathbb{R})$ under the matrix multiplication becomes a group, so it is a subgroup of $GL_n(\mathbb{R})$. We call it the *special linear group* of degree n over $\mathbb{R}$. Its elements are linear transformations under which the volumes are invariant in $\mathbb{R}^n$.

(2) Define

$$O(n) = \{A \in M_{n\times n}(\mathbb{R}) \mid A^T A = I_n\},$$

called the *orthogonal group*, which is the group of linear rigid motions in $\mathbb{R}^n$. Note that this is a subgroup of $GL_n(\mathbb{R})$. We call a matrix $A \in O(n)$ to be an *orthogonal matrix* in the Euclidean space $\mathbb{R}^n$.[17] We also define the *special orthogonal group* of degree n over $\mathbb{R}$,

$$SO(n) := O(n) \cap SL_n(\mathbb{R}) \leq O(n).$$

This is also called the rotation group, whose elements are orientation preserving linear rigid motions in $\mathbb{R}^n$.[18]

EXERCISE 2.6. Let G be a group.

(1) Is a group G abelian if every proper subgroups is abelian?

(2) Show the following: If $H_1 \leq G$ and $H_2 \leq G$, then $H_1 \cap H_2$ is a subgroup of G. In general, if $\{H_i | \ i \in I\}$ is a collection of subgroups of G, then $\bigcap_{i \in I} H_i \leq G$.

(3) Let H and K be subgroups of G. Then we define a set

$$HK = \{hk | \ h \in H, k \in K\}.$$

Show that if G is abelian then HK is a subgroup of G. Note that the smallest subgroup of G containing H and K is denoted by $H \vee K$.

(4) Let a be one fixed element of G. Show that

$$H_a = \{x \in G | \ xa = ax\}$$

is a subgroup of G. The subgroup H_a is called the centralizer of a in G, denote by $Z_G(a)$. Generalizing this definition, show that, for any subset S of G,

$$H_S = \{x \in G | \ xs = as \text{ for all } s \in S\}$$

is a subgroup of G. In particular, the subgroups H_G is called the *center* of G, denoted by $Z(G)$.[19] Show that $Z(G)$ is an abelian group. What is the center $Z(S_3)$ of the symmetric group S_3?

REMARK 2.7. * In an introductory course in algebraic topology[20], we study the fundamental group $\pi_1(S)$ and covering spaces of a space S. All covering spaces of

[17] An orthogonal matrix has determinant ± 1, corresponding to orientation-preserving or orientation-reversing linear rigid motion, respectively.

[18] In fact, every element of $SO(3)$ represents a rotation about an axis in $\mathbb{R}^3$, and the composition of two rotations is a rotation in $\mathbb{R}^3$. See [At, Ch.4 §5].

[19] Following a German, das *Zentrum*.

[20] For very basic materials, see [Ka].

S are determined in term of the subgroups of $\pi_1(S)$. For example, for the circle S^1, its fundamental group is $\pi_1(S^1) = \mathbb{Z}$, and the covering spaces of S^1 are found by the subgroups of $\mathbb{Z}$. In particular, $\pi(\mathbb{R}) = \{0\}$ for the universal covering space $\mathbb{R}$ of S^1.

Let G be a group and $g \in G$. Denote by g^n the element resulting from performing the operation n times. By convention, $g^0 = e$ and $(g^{-1})^{-n}$ for a negative integer n. Note that the subgroup

$$\langle g \rangle = \{g^n | \ n \in \mathbb{Z}\}$$

of G is the smallest subgroup containing the element g, which is called the *cyclic subgroup* generated by $g \in G$.

EXERCISE 2.8. Consider the additive group $\mathbb{Z}_{12} = \{\overline{0}, \overline{1}, \overline{2}, \ldots, \overline{11}\}$ of 12 elements. Compute the subgroups of $\mathbb{Z}_{12}$. Give the subgroup diagram for these subgroups of $\mathbb{Z}_{12}$.

3. Cyclic groups

We study a sort of groups with a simple structure.

DEFINITION 3.1. A group generated by one element is said to be *cyclic*. An element g in a cyclic group $G = \langle g \rangle$ is called a *generator* of G. In general, a set $\{g_1, \ldots, g_k\}$ in a group G is called a generating set (or a set of generators) of G if $\langle g_1, \ldots, g_k \rangle = G$.

The infinite cyclic group $\{\ldots, g^{-2}, g^{-1}, g^0 = e, g, g^2, g^3, \ldots\}$ (with multiplicative notation) has the same structure with $(\mathbb{Z}, +)$ (i.e., *isomorphic* to $(\mathbb{Z}, +)$), which means that there is an *isomorphism*[21] from any infinite cyclic group to $(\mathbb{Z}, +)$.

EXAMPLE 3.2. (1) $\mathbb{Z} = \langle 1 \rangle = \langle -1 \rangle$ with the addition.

(2) $\mathbb{Z}_4 = \{\overline{0}, \overline{1}, \overline{2}, \overline{3}\} = \langle \overline{1} \rangle = \langle \overline{3} \rangle$.

(3) $\mathbb{Z}_{12} = \langle \overline{1} \rangle = \langle \overline{5} \rangle = \langle \overline{7} \rangle = \langle \overline{11} \rangle$.

(4) Neither $(\mathbb{Z}_2 \times \mathbb{Z}_2, +)$ nor $(S_3, \circ)$ is not cyclic.

Proposition 3.3. *Every cyclic group is abelian.*

Proof. Let $G = \langle g \rangle$ be a cyclic group. Then $g^a g^b = g^{a+b} = g^{b+a} = g^b g^a$. □

EXAMPLE 3.4. (1) $\mathbb{Z}$ and $\mathbb{Z}_n$ (n, a positive integer) are clearly abelian.

(2) A non-abelian group, for instance S_3, cannot be a cyclic group.

[21] An isomorphism between two algebraic structures is a bijective map which preserves the operations.

Theorem 3.5. *Every subgroup of a cyclic group is cyclic.*

Proof. Let $H \leq G = \langle g \rangle$ where g is a generator of G. If H is not trivial, take the smallest $m \in \mathbb{Z}_+$ such that $g^m \in H$. We show that g^m is a generator of H by the division algorithm for $\mathbb{Z}$. For any $g^n \in H$ $(n \in \mathbb{Z})$, there exist $q, r \in \mathbb{Z}$ with $0 \leq r < m$ such that $n = qm + r$. Since H is a subgroup, we have $g^r = g^n(g^m)^{-q} \in H$, whence $g^r = g^0 = e$ because it is impossible that $g^r \in H$ for $0 < r < m$. Hence $g^n = (g^m)^q$. □

Remark 3.6. (1) To find all subgroups of a cyclic group, we have only to take the cyclic subgroups.

(2) Consider the group $\mathbb{Z}_2 \times \mathbb{Z}_2 \times \mathbb{Z}_2$ of order 8 and find its subgroups. Among them, the subgroup $\mathbb{Z}_2 \times \mathbb{Z}_2 \times \{\overline{0}\}$ is never cyclic.

Corollary 3.7. *Since $\mathbb{Z}$ is a cyclic group, every subgroup of $\mathbb{Z}$ is of the form $\langle n \rangle = n\mathbb{Z}$ for a nonnegative integer.*

Definition 3.8. Let a, b be positive integers. The positive generator d of the subgroup

$$\langle a, b \rangle = \{ra + sb |\ r, s \in \mathbb{Z}\} \leq \mathbb{Z}$$

under the addition is called the greatest common divisor of a and b, denoted by $\gcd(a, b)$.

Remark 3.9. (1) The above definition for $\gcd(a, b)$ coincides with the definition that d is the greatest among the common divisors of a and b.

First, $d|a$ and $d|b$ from $a \in H := \langle a, b \rangle$ and $b \in H$. Suppose that $dd' = \gcd(a, b)$ for some $d' > 1$ and the unique $\gcd(a, b)$. Then $a = dd'a'$ and $b = dd'b'$ with $\gcd(a', b') = 1$, so $H = \{dd'(ra' + sb') |\ r, s \in \mathbb{Z}\} \subsetneq \{0, \pm d, \pm 2d, \ldots\} = d\mathbb{Z}$, which is a contradiction. Hence the generator d of H is the greatest common divisor of a and b.

(2) Definition 3.8 just shows the existence of the $\gcd(a, b)$, not producing a way of computing $\gcd(a, b)$. Of course, the gcd is gotten easily once we have the prime factorization of each integer. However, the factorization is not easy when the integers are too large. A general method to find $\gcd(a, b)$ is the *Euclidean algorithm.*[22]

[22]Euclidean algorithm for finding g.c.d. of two integers can be applied to any Euclidean domain such as the univariate polynomial ring $\mathbb{Q}[x]$ with rational coefficients. (See Ch.5 §6.)

EXERCISE 3.10. (1) List the elements of the subgroup generated by $\{12, 42\}$ of $\mathbb{Z}$.

(2) Find the positive integer d such that $\langle 24, 15\rangle = \langle d\rangle \leq \mathbb{Z}$, and find $x, y \in \mathbb{Z}$ with $24x + 15y = d$.

Now we state and prove the structure theorem for cyclic groups.

Theorem 3.11. *Let $G = \langle g\rangle = \{g^i | \ i \in \mathbb{Z}\}$ be a cyclic group.*

(a) *If G is infinite, then G is isomorphic to $(\mathbb{Z}, +)$.*
(b) *If G is finite, then G is isomorphic to $(\mathbb{Z}_n, +)$ for a positive integer n.*

Proof. (a) We claim that $g^h \neq g^k$ for any $h \neq k$. Suppose that $g^h = g^k$ for some $h > k$, that is, $g^h \cdot g^{-k} = g^{h-k} = e$. Let $m = \min\{i | \ g^i = e, \ i \in \mathbb{Z}_+\}$. For any $g^n \in G$, $n = qm + r$ with $0 \leq r < m$, so $g^n = g^{qm+r} = (g^m)^q g^r = g^r$. Hence $G = \{g^i | \ i \in \mathbb{Z}\} = \{g^0, g^1, \ldots, g^{m-1}\}$, contradiction to that G is infinite, proving the claim. An infinite cyclic group

$$G = \{\ldots, g^{-2}, g^{-1}, e, g^1, g^2, \ldots\}$$

under the multiplication is isomorphic to $(\mathbb{Z}, +)$.

(b) We must have $g^h = g^k$ for some $h \neq k$. By argument (a), there is a positive integer m such that $g^m = e$. Let n be the smallest such that $g^n = e$. Then

$$G = \{e, g, g^2, \ldots, g^{n-1}\},$$

which is isomorphic to $(\mathbb{Z}_n, +)$. □

REMARK 3.12. Note that $3\mathbb{Z} \leq \mathbb{Z}$ but $3\mathbb{Z} \cong \mathbb{Z}$. An infinite cyclic group has only 2 generators.

We see that the subgroup structure of $\mathbb{Z}_n$ is very simple.

Proposition 3.13. *Let $G = \langle g\rangle$ be a cyclic group of finite order n. A cyclic subgroup H generated by g^s has order $n/\gcd(n, s)$.*

Proof. Let $|H| = m := \min\{i | \ (g^s)^i = e, \ i \in \mathbb{Z}_+\}$. We have $g^{sm} = e$, so sm is a multiple of n. Note that $\gcd(n, s) = \alpha n + \beta s$ for some $\alpha, \beta \in \mathbb{Z}$, i.e.,

$$\alpha\,(n/\gcd(n, s)) + \beta\,(s/\gcd(n, s)) = 1.$$

From $\gcd\left(n/\gcd(n, s), s/\gcd(n, s)\right) = 1$ and $\frac{ms}{n} = \frac{m(s/\gcd(n,s))}{n/\gcd(n,s)} \in \mathbb{Z}$, we obtain that $n/\gcd(n, s)$ divides m. The smallest m must be $n/\gcd(n, s)$. □

Corollary 3.14. *Let $G = \langle g \rangle$ with $|G| = n$. Then $\langle g^s \rangle = G$ if and only if* $\gcd(s, n) =$ 1, *i.e., s is relatively prime to n.* [23]

EXAMPLE 3.15. (1) $\mathbb{Z}_{18} = \langle \bar{1} \rangle = \langle \bar{5} \rangle = \langle \bar{7} \rangle = \langle \overline{11} \rangle = \langle \overline{13} \rangle = \langle \overline{17} \rangle$, and its proper subgroups are:

$$\langle \bar{2} \rangle = \langle \bar{4} \rangle = \langle \bar{8} \rangle = \langle \overline{10} \rangle = \langle \overline{14} \rangle = \langle \overline{16} \rangle, \quad \langle \bar{3} \rangle = \langle \overline{15} \rangle, \quad \langle \bar{6} \rangle = \langle \overline{12} \rangle, \quad \langle \bar{9} \rangle \text{ and } \langle \bar{0} \rangle.$$

(2) Consider the cyclic group $\mathbb{Z}_{48}$. What is the order of the subgroup $\langle \bar{2} \rangle$? What is the order of $\langle \overline{10} \rangle$? The subgroup diagram of $\mathbb{Z}_{48}$ is Figure 1.

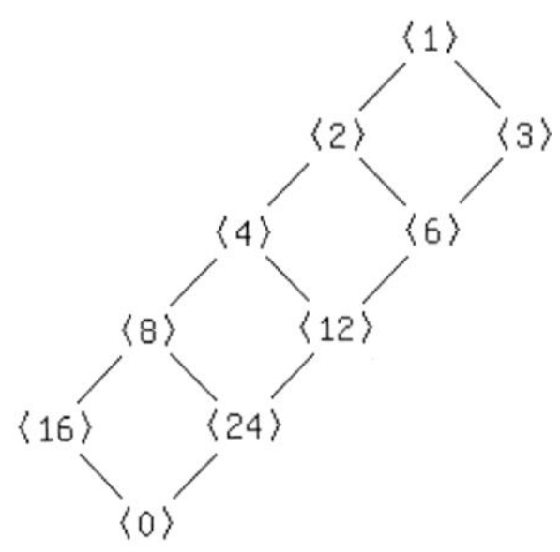

FIGURE 1. The subgroup diagram of $\mathbb{Z}_{48}$

EXERCISE 3.16. (1) Classify all subgroups of $\mathbb{Z}_{60}$.

(2) Let p be a prime number.

1) Find the number of generators of the cyclic group $\mathbb{Z}_{p^r}$, where r is an integer ≥ 1.

2) Show that $\mathbb{Z}_p$ has no proper nontrivial subgroups.

(3) Let G be an abelian group and let H and K be finite cyclic subgroups with $|H| = r$ and $|K| = s$.

1) Show that if r and s are relatively prime, then G contains a cyclic subgroup of order rs.

2) Show that G contains a cyclic subgroup of order the least common multiple of r and s.

REMARK 3.17. (1) In fact, if G is cyclic with a generator g, we have a surjective map $f : \mathbb{Z} \twoheadrightarrow G$ by $f(a) = g^a$. Since every subgroup of $\mathbb{Z}$ is cyclic of the form $n\mathbb{Z}$

[23] The number of integers s such that $1 \leq s < n$ and $\gcd(s, n) = 1$ is denoted by $\phi(n)$, the Euler ϕ-function.

for a nonnegative integer n (by the division algorithm for $\mathbb{Z}$), every subgroup of G is cyclic of the form $f(n\mathbb{Z})$.

(2) The Klein 4-group $V = \{e, a, b, c\}$ is not cyclic. The subgroup $\langle a \rangle$ of V is proper of order 2. But, $\langle a, b\rangle = \langle b, c\rangle = \langle c, a\rangle = V$.

(3) Note that $2\mathbb{Z} \cap 3\mathbb{Z} = 6\mathbb{Z} < \mathbb{Z}$. But $2\mathbb{Z} \cup 3\mathbb{Z}$ is not a subgroup of $\mathbb{Z}$.

We extend the notion of a cyclic group to a group generated by more than one element.

DEFINITION 3.18. Let G be a group and $g_i \in G$ be elements of G. We define

$$\langle \{g_i |\ i \in I\}\rangle \quad (I,\ \text{an index set})$$

to be the smallest subgroup containing g_i's. For a group $G = \langle \{g_i |\ i \in I\}\rangle$, $\{g_i |\ i \in I\}$ is called a set of *generators* of G.

Theorem 3.19. *Let G be a group and $g_i \in G$. Then*

$$\begin{aligned} \langle \{g_i |\ i \in I\}\rangle &= \bigcap \{H |\ H \le G, H \supseteq \{g_i |\ i \in I\}\} \\ &= \left\{ g_{i_1}^{n_1} g_{i_2}^{n_2} \cdots g_{i_k}^{n_k} |\ n_j \in \mathbb{Z}, g_{i_j} \in \{g_i |\ i \in I\} \right\}. \end{aligned}$$

Proof. Let $L = \bigcap \{H |\ H \le G, H \supseteq \{g_i |\ i \in I\}\}$. Since $L \le G$ and $L \supseteq \{g_i |\ i \in I\}$, for all $H \le G$ with $H \supseteq \{g_i |\ i \in I\}$, we have $H \ge L$. That is, L is the smallest subgroup containing $\{g_i |\ i \in I\}$.

For the second equality, let $M = \left\{ g_{i_1}^{n_1} g_{i_2}^{n_2} \cdots g_{i_k}^{n_k} |\ n_j \in \mathbb{Z}, g_{i_j} \in \{g_i |\ i \in I\} \right\}$. It is clear that $M \subseteq \langle \{g_i |\ i \in I\}\rangle$. Now, we note that M satisfies each condition of a group, since $(g_{i_1}^{n_1} \cdots g_{i_k}^{n_k})(g_{s_1}^{l_1} \cdots g_{s_t}^{l_t}) = g_{i_1}^{n_1} \cdots g_{i_{k+t}}^{n_{k+t}} \in M$, $e = g^0 \in M$ and $(g_{i_1}^{n_1} \cdots g_{i_k}^{n_k})^{-1} = g_{i_k}^{-n_k} \cdots g_{i_1}^{-n_1} \in M$. So $M \le G$, and together with $M \supseteq \{g_i |\ i \in I\}$, we get that $M \ge \langle \{g_i |\ i \in I\}\rangle$. □

4. Dihedral groups

We are invited to have a series of non-abelian groups.

DEFINITION 4.1. Let A be a set. A function $f : A \xrightarrow[\text{onto}]{1-1} A$ is called a *permutation* of A.

REMARK 4.2. (1) Let $S_A = \{f|\ f : A \to A, \text{ a permutation}\}$. Then $(S_A, \circ)$ is a group.

(2) In particular, if $A = \{1, 2, \ldots, n\}$ is a finite set, then $S_n := S_A$ is called the nth *symmetric group*. Note that $|S_n| = n!$ and $S_n \leq GL_n(\mathbb{R})$. For example,

$$S_3 \longleftrightarrow \left\{ \begin{pmatrix} 1 & 0 & 0 \\ 0 & 1 & 0 \\ 0 & 0 & 1 \end{pmatrix}, \begin{pmatrix} 0 & 1 & 0 \\ 0 & 0 & 1 \\ 1 & 0 & 0 \end{pmatrix}, \begin{pmatrix} 0 & 0 & 1 \\ 1 & 0 & 0 \\ 0 & 1 & 0 \end{pmatrix}, \begin{pmatrix} 0 & 1 & 0 \\ 1 & 0 & 0 \\ 0 & 0 & 1 \end{pmatrix}, \begin{pmatrix} 1 & 0 & 0 \\ 0 & 0 & 1 \\ 0 & 1 & 0 \end{pmatrix}, \begin{pmatrix} 0 & 0 & 1 \\ 0 & 1 & 0 \\ 1 & 0 & 0 \end{pmatrix} \right\}.$$

EXAMPLE 4.3 ($S_3 = D_{2\cdot 3}$). Let

$$S_3 = \{1, (1,2,3), (1,3,2), (1,2), (2,3), (1,3)\},$$

which is equal to $D_{2\cdot 3}$, the group of symmetries of a regular triangle.

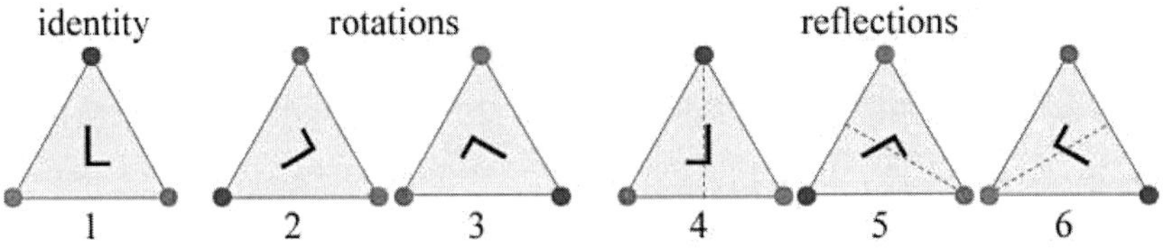

FIGURE 2. 6 elements of $D_{2\cdot 3}$

Note that S_3 is not abelian because $(1,2) \circ (1,2,3) \neq (1,2,3) \circ (1,2)$. We know that

$$(1,2) \circ (1,2,3) = (1,3,2) \circ (1,2).$$

The nontrivial proper subgroups of S_3 are

$$\{1, (1,2)\},\ \{1, (2,3)\},\ \{1, (1,3)\} \text{ and } \{1, (1,2,3), (1,3,2)\},$$

which are all abelian.

REMARK 4.4. (1) A *Cayley*[24] *digraph* of a group G is a directed graph consisting of vertices and arcs between two vertices, where each vertex represents an element in G and each directed edge represents the multiplication by an element in G from the right. Draw a Cayley digraph of $S_3 = D_{2\cdot 3}$.

(2) Note that S_3 is the smallest nonabelian group. We will see that every group of order less than 6 is abelian, and that, in general, if a nonabelian group G is of order $2p$ where p is an odd prime then G is isomorphic to the dihedral group D_{2p}.

EXAMPLE 4.5 ($S_4 > D_{2\cdot 4}$). (1) The group S_4 is of order $4! = 24$ and is not abelian since S_3 is not. Note that S_4 is the group of orientation-preserving symmetries of a cube(or a regular octahedron).

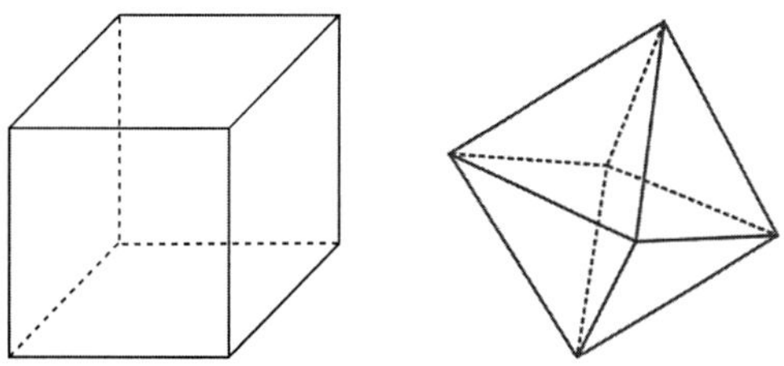

FIGURE 3. A cube has 4 pairs of diagonal vertices, and a regular octahedron has 4 pairs of diagonal faces.

(2) Consider a proper subgroup of S_4,

$$D_{2\cdot 4} = \{1, (1,2,3,4), (1,3)(2,4), (1,4,3,2), (1,3), (2,4), (1,4)(2,3), (1,2)(3,4)\},$$

the group of symmetries of a square, which we call the 4th *dihedral group*. If we let $a = (1,2,3,4)$ and $b = (1,3)$ then we know that

$$a^4 = e,\ b^2 = e \ \text{ and } ab = ba^3 = ba^{-1}.$$

REMARK 4.6. * In general, the nth dihedral group is the group of symmetries of a regular n-polygon,

$$D_{2n} = \{1, \rho, \rho^2, \ldots, \rho^{n-1}, \mu_1, \ldots, \mu_n\},$$

where ρ is the rotation by $2\pi/n$ and μ_i's are reflections. Note that $|D_{2n}| = 2n$ and

$$\rho^n = 1,\ \mu^2 = e \ \text{ and } \rho\mu = \mu\rho^{-1},$$

[24]Arthur Cayley(1821–1895), a British mathematician. He was the first to define the concept of a *group* in the modern way. Before he became a professor at Cambridge, he worked as a lawyer for 14 years. While teaching at Cambridge, he was instrumental in persuading the administration to begin accepting women as students.[LF]

where $\mu = \mu_i$ for each $i = 1, \ldots, n$. [25]

EXERCISE 4.7. Draw a Cayley digraph of D_{2n}, and find all subgroups of D_{2n} when $n = 4$.

EXERCISE 4.8. * An element of order 2 in a group is called an *involution*. The structure of a finite group can be obtained by studying its involutions and their centralizers. Let u and v be distinct involutions in a finite group G. Show the following.

(1) $\langle u, v\rangle$ is isomorphic to a dihedral group. [Hint: Consider a presentation.]

(2) If uv is of odd order then u and v are conjugate in G, i.e., there exists $a \in G$ such that $v = aua^{-1}$.

(3) If uv is of even order $2n$ so $w = (uv)^n$ is an involution, then $u, v \in Z_G(w)$ where $Z_G(w) = \{x \in G |\ xw = wx\}$ is the centralizer of w in G.

[25] In fact, these define the dihedral group D_{2n}, that is, these are the relations in a presentation of D_{2n}.

5. Alternating groups

We already observed in the course of Linear Algebra that the groups

$$S_n = \{\text{premutations of } \{1, 2, \dots, n\}\}$$

and A_n arise in the definition of determinants of square matrices.

DEFINITION 5.1. For a fixed permutation σ of a set S, and $a, b \in S$, we denote by $a \sim b$ if $b = \sigma^n(a)$ for some $n \in \mathbb{Z}$. Then $\sim$ is an equivalence relation, i.e., it is reflexive, symmetric and transitive. We define

$$[a] := \{b|\ a \sim b\} = \{\sigma^n(a)|\ n \in \mathbb{Z}\},$$

which is called the *orbit* of σ containing a.

EXERCISE 5.2. Explain why we can identify an *equivalence relation* with a *partition*.

Let m be the least positive integer such that $\sigma^m(a) = a$. Then the set

$$[a] = \{\sigma^i(a)|\ 0 \le i \le m-1\}$$

is finite. Note that $b \in [a]$ if and only if $[a] = [b] = \{\sigma^i(b)|\ 0 \le i \le m-1\}$.

DEFINITION 5.3. A permutation σ on S is called a *cycle* if σ has at most one orbit which has more than one element, that is, $\sigma = \text{id}$ or $\sigma = \big(a, \sigma(a), \dots, \sigma^{m-1}(a)\big)$ for some $a \in S$ and a positive integer m. The length of a cycle is the number of elements in its largest orbit.

EXAMPLE 5.4. Let $\sigma = \begin{pmatrix} 1 & 2 & 3 & 4 & 5 & 6 & 7 & 8 \\ 3 & 8 & 6 & 7 & 4 & 1 & 5 & 2 \end{pmatrix} \in S_8$. Then the orbits of σ are $\{1, 3, 6\}$, $\{2, 8\}$ and $\{4, 5, 7\}$, and the permutation σ is expressed by $\sigma = (1, 3, 6)(2, 8)(4, 7, 5)$.

Theorem 5.5. *Every permutation σ of a finite set is a product of disjoint cycles.*

Proof. We know that an equivalence relation induces a partition. Let B_i $(i = 1, 2, \dots, r)$ be disjoint orbits of σ. Define the cycles σ_i $(i = 1, 2, \dots, r)$ by

$$\sigma_i(x) = \begin{cases} \sigma(x) & \text{for } x \in B_i; \\ x & \text{otherwise.} \end{cases}$$

Then $\sigma = \sigma_1\sigma_2 \cdots \sigma_r$ where σ_i's are disjoint cycles. □

EXERCISE 5.6. (1) Express the permutation $\begin{pmatrix} 1 & 2 & 3 & 4 & 5 & 6 & 7 & 8 \\ 8 & 2 & 6 & 3 & 7 & 4 & 5 & 1 \end{pmatrix}$ as a product of disjoint cycles.

(2) Express the product $\sigma = (1,4,5)(7,8)(2,5,7)$ of cycles as a product of disjoint cycles. What is the order of the subgroup $\langle\sigma\rangle$ of S_8?

(3) Let $H = \{\sigma \in S_4 |\ \sigma \text{ has 2 or 4 orbits}\}$. How many elements are in H? Is H a subgroup of S_4?

(4) Show that if σ is a cycle of odd length then σ^2 is a cycle.

(5) Find the maximal possible order for an element of S_n for $n = 10$.

(6) Let $\tau = (1,2,4,3)(5,6)$ and $\sigma = (1,3,4)(2,6)$. Compute σ^{-1} and $\sigma\tau\sigma^{-1}$.

(7)* A permutation σ can be written as a product of disjoint cycles, and let $d_1, \ldots, d_r$ be the number of elements of each cycle in increasing order. Let τ be another permutation which can be written as a product of disjoint cycles whose cardinalities are $d'_1, \ldots, d'_s$ in increasing order. Prove that σ is conjugate to τ in S_n if and only if $r = s$ and $d_i = d'_i$ for all $i = 1, \ldots, r$, i.e., two permutations have the same cycle type. [Hint: If $c = (i_1 \cdots i_m)$ is a cycle and $\gamma \in S_n$, then $\gamma c \gamma^{-1}$ is the cycle $(\gamma(i_1) \cdots \gamma(i_m))$.]

DEFINITION 5.7. A *transposition* is a cycle of length 2.

A computation shows that a cycle is a composition of transpositions as follows:

$$\begin{aligned}(a_1, a_2, \ldots, a_k) &= (a_1, a_k) \circ (a_1, a_{k-1}) \circ \cdots \circ (a_1, a_3) \circ (a_1, a_2) \\ &= (a_1, a_2)(a_1, a_3) \cdots (a_1, a_{k-1})(a_1, a_k).\end{aligned}$$

Note that the first expression follows the left notation, on the other hand, the second is as the *right* notation, which is natural.

Corollary 5.8. *Any permutation in S_n is a product of transpositions.*

DEFINITION 5.9. A transposition of the form $(i, i+1) \in S_n$ for $i = 1, \ldots, n-1$ is called an *adjacent transposition.*

EXERCISE 5.10. (1) Show that S_n is generated by $n-1$ transpositions[26]

$$(1,2), (1,3), \ldots, (1,n).$$

(2) Show that S_n is generated by the adjacent transpositions

$$(1,2), (2,3), \ldots, (n-1,n).$$

[26] $(a_1, a_2, \ldots, a_k) = (1, a_1)(1, a_2) \cdots (1, a_k)(1, a_1)$

[Hint: Every transposition is a product of adjacent transpositions.] [27]

(3) Show that S_n is generated by $\{(1,2),(1,2,3,\ldots,n)\}$.

[Hint: Consider $(1,2,3,\ldots,n)^r(1,2)(1,2,3,\ldots,n)^{n-r}$ for $r=0,1,\ldots,n-2$.]

REMARK 5.11. * The group S_n is a reflection group, i.e., a group generated by reflections, elements of order 2. Note that S_n arises as the Weyl[28] group of the root system of type A_{n-1} in Lie[29] theory.

Theorem 5.12 (Cayley). * *Every group is isomorphic to a subgroup of S_n for some $n \in \mathbb{Z}_+ \cup \{\infty\}$.*

Proof. Let G be a group. Then we want to show that there exists a group G' isomorphic to G such that $G' < S_G = \{\sigma |\ \sigma : G \xrightarrow[\text{onto}]{1-1} G\}$. Define a map

$$\lambda : G \to S_G \text{ by } a \mapsto \lambda_a, \quad \text{where } \lambda_a(x) = ax$$

for each $x \in G$. Since $ax = ay$ implies $x = y$, each λ_a $(a \in G)$ is 1-1. For each $b \in G$, $\lambda_a(a^{-1}b) = b$, which means that λ_a $(a \in G)$ is onto. So the map λ is well-defined. Furthermore, it is 1-1 because if $\lambda_a = \lambda_b$, i.e., $\lambda_a(x) = \lambda_b(x)$ for each $x \in G$ then $a = b$. Now we claim that $G' := \operatorname{Im}\lambda = \{\lambda_a |\ a \in G\}$ is a subgroup of S_G. Since $\lambda_a \circ \lambda_b(x) = abx = \lambda_{ab}(x)$ for $x \in G$, G' is closed under the composition. The identity element is $\lambda_e : G \to G$ (e is the identity in G), and $(\lambda_a)^{-1} = \lambda_{a^{-1}}$ for $a \in G$. So $G' < S_G$. By restricting the codomain of λ, we have an isomorphism between G and $G' = \operatorname{Im}\lambda$. □

REMARK 5.13. * We may consider a map

$$\rho : G \to S_G \text{ by } a \mapsto \rho_a, \text{ where } \rho_a(x) = xa^{-1}$$

for each $x \in G$. Then G is isomorphic to $G'' = \operatorname{Im}\rho < S_G$. We say that λ is the left regular representation of G and ρ is the right regular representation of G.

Now we define the alternating groups.

Theorem 5.14. *Let $\sigma \in S_n$ be a permutation. If $\sigma = \tau_1\tau_2\cdots\tau_s = \eta_1\eta_2\cdots\eta_t$ where τ_i's and η_j's are transpositions, then $s - t$ is even.*

---

[27] $(i,j) = (i,i+1)(i+1,i+2)\cdots(j-1,j)(j-2,j-1)\cdots(i,i+1)$

[28] Hermann Weyl (1885–1955), a German mathematician.

[29] Sophus Lie (1842–1899), a Norwegian mathematician.

Lemma 5.15. *Let $\sigma \in S_n$ be a permutation and $\tau = (i, j) \in S_n$ be a transposition. Then*

$$|\text{the number of orbits of } \sigma\tau - \text{the number of orbits of } \sigma| = 1.$$

Proof. We divide the situations into two cases: Case 1) i and j are in different orbits; and Case 2) i and j are in the same orbit of σ.

Case 1) Let $\sigma = (u, \ldots, a, i, \ldots, v)(x, \ldots, b, j, \ldots, y)\sigma_3 \cdots \sigma_r$ be a product of disjoint cycles, i.e., $\sigma_1 = (u, \ldots, a, i, \ldots, v)$, $\sigma_2 = (x, \ldots, b, j, \ldots, y)$. Then $\tau \circ (\sigma_1\sigma_2)$ becomes one cycle. In other words, two orbits of σ are joined into a single orbit when we consider $\tau \circ \sigma$.

Case 2) Let $\sigma = (x, \ldots, a, i, \ldots, b, j, \ldots, y)\sigma_2 \cdots \sigma_r$ be a product of disjoint cycles, i.e., $\sigma_1 = (x, \ldots, a, i, \ldots, b, j, \ldots, y)$. Then $\tau \circ \sigma_1$ becomes a product of two cycles. That is, the orbit of σ_1 is split into two orbits. □

Proof of Theorem 5.14. Suppose that s is even and t is odd. Note that $\sigma^{-1} = (\eta_1\eta_2 \cdots \eta_t)^{-1} = \eta_t^{-1} \cdots \eta_2^{-1}\eta_1^{-1} = \eta_t \cdots \eta_2\eta_1$. Since $\sigma\sigma^{-1} = \text{id}$, we compare the numbers of orbits in both sides of

$$(\tau_1\tau_2 \cdots \tau_s)(\eta_t \cdots \eta_2\eta_1) \circ \text{id} = \text{id},$$

thus obtain that the difference of orbits of both sides is odd by Lemma 5.15, which is a contradiction to the equality. □

By Theorem 5.14, it is well defined that a permutation $\sigma \in S_n$ is *even*(or *odd*) if it is written as a product of even(odd, respectively) number of transpositions. Let A_n be the set of all even permutations of S_n, and B_n be the set of all odd permutations of S_n. Then we note that A_n is a group, and $S_n = A_n \amalg B_n$ with $|A_n| = |B_n| = |S_n|/2 = n!/2$ since the map $A_n \to B_n$ by $\sigma \mapsto \sigma(1, 2)$ is a bijection.

Definition 5.16. The subgroup

$$A_n := \{\text{even permutations in } S_n\}$$

is called the nth *alternating group.* [30]

---

[30]We will see that A_n is a simple nonabelian group if $n \geq 5$. This is an essence of a solution to why an algebraic equation of degree ≥ 5 is not solvable by radicals. It was Abel(1802–1829, a Norwegian mathematician) who proved the insolvability of a general quintic equation, an amazing

REMARK 5.17. (1) We define sign : $S_n \to \{\pm 1\}$ by $\mathrm{sign}(\sigma) = 1$ ($\mathrm{sign}(\sigma) = -1$) if σ is an even(odd, respectively) permutation. Then this is a group homomorphism[31] with $\ker(\mathrm{sign}) = A_n$.

(2) Let $A = (a_{ij})$ be an $n \times n$ matrix with real entries. Then

$$\det A = \sum_{\sigma \in S_n} \mathrm{sign}(\sigma) a_{1\sigma(1)} a_{2\sigma(2)} \cdots a_{n\sigma(n)}.$$

EXERCISE 5.18. (1) Show that a permutation with odd order must be an even permutation.

(2) Show that for every subgroup H of S_n for $n \geq 2$, either all permutations in H are even or exactly half of them are even.

EXERCISE 5.19. Consider the following combination of numbers in a square with one blank. By moving the numbers using the blank, is it possible to relocate the numbers in a right order with finally the same position of the blank?

15	14	13	12
11	10	9	8
7	6	5	4
3	2	1	

We close this section with various examples: the groups of symmetries of a regular polyhedron.

EXERCISE 5.20. (1) Write down all elements of the alternating group A_4 and find a subgroup of order 4. Draw a Cayley digraph for A_4, using the generating set $\{(1,2,3),(1,2)(3,4)\}$. [Answer: Figure 4.][32]

(2) Prove that the group of rotational symmetries of a regular tetrahedron is A_4. [Hint: Figure 5.]

EXERCISE 5.21. Prove that:

(1) The full group of symmetries(rotations and reflections) of a regular tetrahedron is S_4.[33]

fact. Évariste Galois(1811–1832, a French mathematician) discovered the notion of a solvable group. He was the first to realize the importance of the existence of normal subgroups.

[31] A homomorphism is a map which preserves the operations.

[32] See [F, Figure 9.22].

[33] `http://en.wikipedia.org/wiki/Tetrahedral_symmetry`.

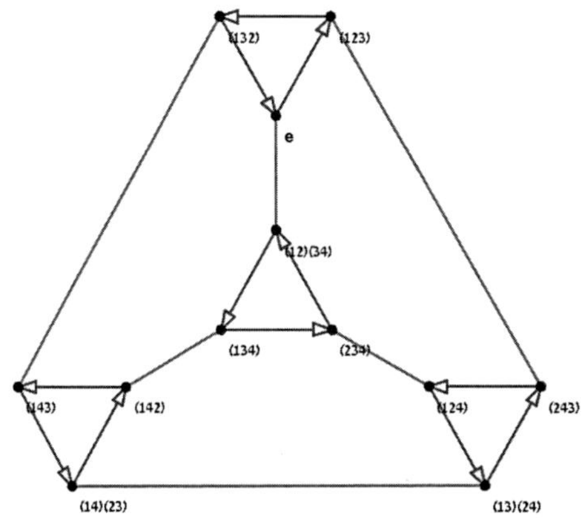

FIGURE 4. Cayley digraph of A_4

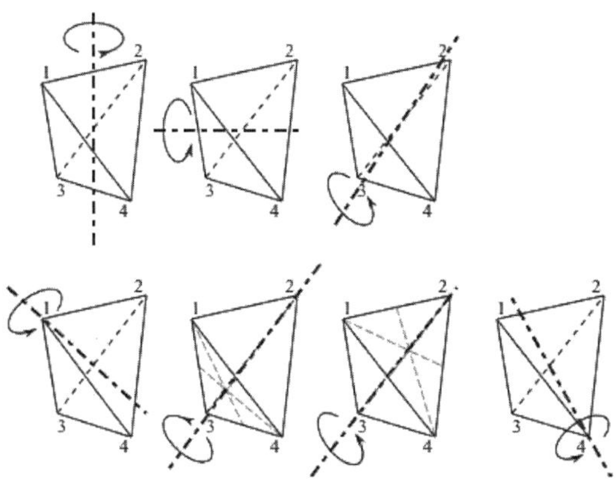

FIGURE 5. Rotations of a regular tetrahedron

(2) The group of rotational symmetries of a cube or of a regular octahedron is S_4.[34]

(3)* The full group of symmetries of a cube or a regular octahedron is $S_3 \ltimes \mathbb{Z}^3$.

(4) The group of rotational symmetries of a regular icosahedron or of a regular dodecahedron is A_5.[35]

[34] http://en.wikipedia.org/wiki/Octahedral_symmetry.

[35] http://en.wikipedia.org/wiki/Icosahedral_symmetry.

REMARK 5.22. * (1) The full group of symmetries of a regular icosahedron is $A_5 \times \mathbb{Z}_2$ of order 120, not isomorphic to S_5. In general, the full icosahedral group is among the finite Coxeter groups.[36] It is the Coxeter group of type H_3.

(2) There are 6 kinds of convex regular 4-dimensional polytopes:[37]

4-simplex(5-cell), 4-cube(8-cell), 4-orthoplex(4-cross-polytope, 16-cell),
24-cell, 120-cell, 600-cell.

The corresponding symmetry groups are:

$$A_4 (= S_5),\ B_4 (= S_4 \ltimes \mathbb{Z}^4),\ B_4,\ F_4,\ H_4,\ H_4.$$

(3) In dimensions $n \geq 5$, there are only three kinds of regular polytopes:

n-simplex, n-cube, n-orthoplex.

The corresponding symmetry groups are:

$$A_n,\ B_n,\ B_n.$$

(4) Note that the Coxeter groups of type

$$D_n,\ E_6,\ E_7,\ E_8$$

are the symmetry groups of certain semiregular polytopes.

---

[36] Harold Scott MacDonald Coxeter(1907–2003), a British-born Canadian geometer. For a general accounts for Coxeter groups, see [H90].

[37] `http://en.wikipedia.org/wiki/Convex_regular_polychoron`.

6. Finitely generated abelian groups

Let $S_1, S_2, \ldots, S_n$ be arbitrary sets. We define the *Cartesian product* of S_i's to be

$$\prod_{i=1}^{n} S_i = S_1 \times S_2 \times \cdots \times S_n = \{(a_1, a_2, \ldots, a_n) |\ a_i \in S_i, i = 1, 2, \ldots, n\}.$$

REMARK 6.1. * For an infinite index set I,

$$\prod_{i\in I} S_i := \{f : I \to \cup_{i\in I} S_i |\ f(i) \in S_i\}.$$

Proposition 6.2. *Let $G_1, \ldots, G_n$ be groups. Then $\prod_{i=1}^{n} G_i$ is a group by defining the group operation componentwise, i.e.,*

$$(a_1, a_2, \ldots, a_n)(b_1, b_2, \ldots, b_n) = (a_1b_1, a_2b_2, \ldots, a_nb_n) \in \prod_{i=1}^{n} G_i.$$

We call $\prod_{i=1}^{n} G_i$ the *direct product* of the groups G_i's. Usually, for abelian groups G_i $(i = 1, \ldots, n)$, we write $\bigoplus_{i=1}^{n} G_i = G_1 \oplus \cdots \oplus G_n$ in place of $\prod_{i=1}^{n} G_i$.

REMARK 6.3. * Let I be an index set (possibly infinite). The *direct product*

$$\prod_{i\in I} G_i = \{f : I \to \cup_{i\in I} G_i |\ f(i) \in G_i\}$$

is a group endowed with $ab : I \to \cup_{i\in I} G_i$ where

$$(ab)(i) := a(i)b(i) \in G_i$$

for $a, b \in \prod_{i\in I} G_i$. The *direct sum* of G_i's is defined by

$$\bigoplus_{i\in I} G_i := \{f : I \to \cup_{i\in I} G_i |\ f(i) \in G_i,\ f(i) = e \in G_i \text{ except for finitely many } i\text{'s}\}.$$

Note that for a finite set I, the direct sum is the same as the direct product, however, for an infinite index set, the direct sum is properly contained in the direct product.

EXAMPLE 6.4. (1) $\mathbb{Z}_2 \times \mathbb{Z}_2$ is a group of order 4, but is not cyclic since any element is of order 2. Note that $\mathbb{Z}_2 \times \mathbb{Z}_2$ is isomorphic to the Klein 4-group.

(2) $\mathbb{Z}_2 \times \mathbb{Z}_3$ is a cyclic group of order 6 with generator $(\bar{1}, \bar{1})$. So $\mathbb{Z}_2 \times \mathbb{Z}_3$ is isomorphic to $\mathbb{Z}_6$.

Theorem 6.5. *The group $\mathbb{Z}_m \times \mathbb{Z}_n$ is cyclic if and only if* $\gcd(m, n) = 1$.

Proof. $\Leftarrow$) Take $(\overline{1},\overline{1}) \in \mathbb{Z}_m \times \mathbb{Z}_n$. Consider the subgroup $\langle(\overline{1},\overline{1})\rangle = \{k(\overline{1},\overline{1}) \mid k \in \mathbb{Z}\}$. The least positive integer k such that $k(\overline{1},\overline{1}) = (\overline{k},\overline{k}) = (\overline{0},\overline{0}) \in \mathbb{Z}_m \times \mathbb{Z}_n$ is $k = \operatorname{lcm}(m,n)$. So $|\langle(\overline{1},\overline{1})\rangle| = \operatorname{lcm}(m,n) = mn = |\mathbb{Z}_m \times \mathbb{Z}_n|$, that is, $\langle(\overline{1},\overline{1})\rangle = \mathbb{Z}_m \times \mathbb{Z}_n$.

$\Rightarrow$) Suppose that $\gcd(m,n) = d > 1$. Take any $(\overline{a},\overline{b}) \in \mathbb{Z}_m \times \mathbb{Z}_n$ and consider the subgroup $\langle(\overline{a},\overline{b})\rangle = \{k(\overline{a},\overline{b}) \mid k \in \mathbb{Z}\}$. Then there exists a positive integer $k = \operatorname{lcm}(m,n) = mn/d < mn$ such that $k(\overline{a},\overline{b}) = (\overline{ka},\overline{kb}) = (\overline{0},\overline{0}) \in \mathbb{Z}_m \times \mathbb{Z}_n$. Thus, no element $(\overline{a},\overline{b})$ generates $\mathbb{Z}_m \times \mathbb{Z}_n$, which contradicts that $\mathbb{Z}_m \times \mathbb{Z}_n$ is cyclic. $\square$

Corollary 6.6. (a) *The group* $\mathbb{Z}_{m_1} \times \mathbb{Z}_{m_2} \times \cdots \times \mathbb{Z}_{m_n}$ *is cyclic if and only if* $\gcd(m_i, m_j) = 1$ *for* $i \neq j$.

(b) *The group* $\mathbb{Z}_{(p_1)^{e_1}} \times \mathbb{Z}_{(p_2)^{e_2}} \times \cdots \times \mathbb{Z}_{(p_r)^{e_r}}$ *(p_i's are distinct primes) is cyclic of order* $(p_1)^{e_1}(p_2)^{e_2}\cdots(p_r)^{e_r}$.

Example 6.7. $\mathbb{Z}_8 \times \mathbb{Z}_9 \cong \mathbb{Z}_{72}$. $\mathbb{Z}_2 \times \mathbb{Z}_2 \times \mathbb{Z}_2 \not\cong \mathbb{Z}_2 \times \mathbb{Z}_4 \not\cong \mathbb{Z}_8$. $\mathbb{Z}_3 \times \mathbb{Z}_3 \not\cong \mathbb{Z}_9$.

Exercise 6.8. (1) Find the order of $(\overline{8},\overline{10})$ in $\mathbb{Z}_{12} \times \mathbb{Z}_{18}$, and find the order of $(\overline{3},\overline{10},\overline{9})$ in $\mathbb{Z}_4 \times \mathbb{Z}_{12} \times \mathbb{Z}_{15}$.

(2) What is the largest order among the orders of all cyclic subgroups of $\mathbb{Z}_6 \times \mathbb{Z}_8$?

(3) For $\mathbb{Z}_8$, $\mathbb{Z}_4 \times \mathbb{Z}_2$, $\mathbb{Z}_2 \times \mathbb{Z}_2 \times \mathbb{Z}_2$, determine the number of subgroups of order 4.

(4) Find all subgroups of $\mathbb{Z}_2 \times \mathbb{Z}_2 \times \mathbb{Z}_4$ that are isomorphic to the Klein 4-group. [Hint: First, find every element in $\mathbb{Z}_2 \times \mathbb{Z}_2 \times \mathbb{Z}_4$ whose order is 2.]

We state the structure theorem for finitely generated abelian groups, whose proof needs some knowledge of advanced Linear Algebra.

Theorem 6.9. *Let G be a finitely generated abelian group. Then G is a direct sum of finite cyclic groups and a free abelian group. That is, there is an isomorphism*

$$\phi : G \longrightarrow \mathbb{Z}_{d_1} \times \cdots \times \mathbb{Z}_{d_k} \times \mathbb{Z}^r,$$

where the Betti number r is nonnegative, invariant factors $d_1, \ldots, d_k$ are integers $\neq 0, \pm 1$ and $d_1 | d_2 | \cdots | d_k$. Equivalently, G is isomorphic to

$$\mathbb{Z}_{p_1^{e_1}} \times \cdots \times \mathbb{Z}_{p_n^{e_n}} \times \mathbb{Z}^r,$$

where each elementary divisor $p_1^{e_1}, \ldots, p_n^{e_n}$ is a power of a prime element of $\mathbb{Z}$ with repetitions of primes allowed.

EXAMPLE 6.10. Let us classify all finite abelian groups of order $360 = 2^3 \cdot 3^2 \cdot 5$, up to isomorphism. By the structure theorem, an abelian group of order 360 is isomorphic to one of the following 6 possibilities:

$$\mathbb{Z}_2 \times \mathbb{Z}_2 \times \mathbb{Z}_2 \times \mathbb{Z}_3 \times \mathbb{Z}_3 \times \mathbb{Z}_5, \quad \mathbb{Z}_2 \times \mathbb{Z}_4 \times \mathbb{Z}_3 \times \mathbb{Z}_3 \times \mathbb{Z}_5, \quad \mathbb{Z}_2 \times \mathbb{Z}_2 \times \mathbb{Z}_2 \times \mathbb{Z}_9 \times \mathbb{Z}_5,$$
$$\mathbb{Z}_2 \times \mathbb{Z}_4 \times \mathbb{Z}_9 \times \mathbb{Z}_5, \quad \mathbb{Z}_8 \times \mathbb{Z}_3 \times \mathbb{Z}_3 \times \mathbb{Z}_5, \quad \mathbb{Z}_8 \times \mathbb{Z}_9 \times \mathbb{Z}_5.$$

The number of possible abelian groups of a given order is determined by the product of partition numbers $p(e_i)$ of exponents e_i in the prime factorization of the order. In this example, $p(3) = 3$ and $p(2) = 2$ because $3 = 2+1 = 1+1+1$ and $2 = 1+1$.

Note that the *partition number* $p(n)$ counts the number of abelian groups of order p^n (p, a prime, n, a positive integer), up to isomorphism.

EXERCISE 6.11. (1) Classify all finite abelian groups of order 720, up to isomorphism.

(2) Find the partition number $p(5)$. What is $p(6)$?

EXERCISE 6.12. (1) Show that a finite abelian group is not cyclic if and only if it contains a subgroup isomorphic to $\mathbb{Z}_p \times \mathbb{Z}_p$ for some prime p.

(2) Let G, H and K be finitely generated abelian groups. Show that $G \times H \cong G \times K$ implies $H \cong K$.

DEFINITION 6.13. A group G is *decomposable* if $G \cong H_1 \times H_2$ for $\{e\} < H_i < G$ $(i = 1, 2)$. We say that G is *indecomposable* if it not decomposable.

Proposition 6.14. (a) *A finite abelian group G is indecomposable if and only if $G \cong \mathbb{Z}_{p^e}$ for a prime p.*

(b) *Let G be a finite abelian group. If m is a divisor of $|G|$ then G has a subgroup H with $|H| = m$.*

(c) *If G is a finite abelian group G whose order $|G|$ is square free, i.e., $|G|$ is not divisible by the square of any prime, then G is cyclic.*

Proof. Left as exercises. □

Midterm Exam

(October 20, 2011)

1. Determine whether each statement is true or not: True(T) or False(F).
   (If you answer correctly, you get 3 points; If nothing, 1 point; If wrong, 0 point.)
   1) $(\mathbb{Z}_2 \times \mathbb{Z}_2, +)$ is a cyclic group.
   2) An infinite cyclic group has only 2 generators.
   3) $(S_3, \circ)$ is an abelian group.
   4) The set of even permutations of S_n is a group under the composition.
   5) $(\mathbb{Z}_9, +)$ is isomorphic to $(\mathbb{Z}_3 \times \mathbb{Z}_3, +)$ as a group.
2. (15pts) Divide the following 8 groups into classes of isomorphic groups, in other words, determine which groups are isomorphic one another, or not:

   $(\mathbb{Z}_6, +)$, $(\mathbb{Z}_2 \times \mathbb{Z}_3, +)$, the dihedral group D of a regular triangle,
   $(S_3, \circ)$, the group S' of permutations of $\{1, 2, 3, 4\}$ with the number 4 fixed,
   $(\mathbb{Z}_8, +)$, $(\mathbb{Z}_2 \times \mathbb{Z}_4, +)$, the dihedral group D' of a square.

- For each of the following problems, you should justify your answer.

3. (10pts) We define a relation $\sim$ on $\mathbb{Z}$ by $x \sim y$ if $x - y$ is a multiple of 5. Determine whether $\sim$ is an equivalence relation or not, and if so, find its equivalence classes.
4. (10pts) Let G be a set with a binary operation $*$ on G, satisfying that $*$ is associative and there exists $e \in G$ such that $a * e = a = e * a$ for each $a \in G$. Write the additional condition for G to be a group.
5. (10pts) Note that the 4th dihedral group D_4 is the group of symmetries of a square. Find all subgroups of order 2 in D_4.
6. (15pts) (1) Find the positive integer d such that $\langle 142, 90\rangle = \langle d\rangle \leq \mathbb{Z}$, and also find $x, y \in \mathbb{Z}$ such that $142x + 90y = d$.
   (2) What is the number of generators of $(\mathbb{Z}_{25}, +)$?
7. (10pts) Let $\sigma = \begin{pmatrix} 1 & 2 & 3 & 4 & 5 & 6 & 7 & 8 \\ 8 & 2 & 6 & 3 & 7 & 4 & 5 & 1 \end{pmatrix} \in S_8$.
   (1) Express the permutation σ as a product of disjoint cycles.
   (2) What is the order of the subgroup $\langle \sigma \rangle$ of S_8?
8. (15pts) Classify all finite abelian groups of order 288, up to isomorphism.

Midterm Exam

(October 21, 2010)

1. Determine whether each statement is true or not: True(T) or False(F).

   (If you answer correctly, you get 3 points; If nothing, 1 point; If wrong, 0 point.)

   1) An equivalence relation on a set S is the same notion as a partition on S.
   2) The set $H = \left\{\left(\begin{smallmatrix} a & -b \\ b & a \end{smallmatrix}\right) \mid a, b \in \mathbb{R}\right\}$ with the matrix addition and multiplication is isomorphic to the set of complex numbers $(\mathbb{C}, +, \cdot)$.
   3) The subgroup $3\mathbb{Z} = \{3k \mid k \in \mathbb{Z}\}$ of $\mathbb{Z}$ under the addition is not isomorphic to $\mathbb{Z}$.
   4) If a group $(G, *)$ with identity e satisfies that $x * x = e$ for all $x \in G$, then G is abelian.
   5) The set of $n \times n$ invertible matrices with entries in $\mathbb{R}$ is a group under the addition.
   6) The set of $n \times n$ real matrices with determinant 1 is a subgroup of $GL_n(\mathbb{R})$.
   7) The group $\mathbb{Z}_2 \times \mathbb{Z}_2$ under the componentwise addition is generated by one element.
   8) Every subgroup of a cyclic group is cyclic.
   9) The additive group $\mathbb{Z}_2 \times \mathbb{Z}_3$ is cyclic.
   10) The additive group $\mathbb{Z}_2 \times \mathbb{Z}_4$ is isomorphic to $\mathbb{Z}_8$.

- For each of the following problems, you should justify your answer.

2. (20 points) Determine whether each $\sim$ is an equivalence relation or not.

   (1) A relation $\sim$ on $\mathbb{Z}$ is defined by $x \sim y \Leftrightarrow x - y$ is an odd integer.

   (2) A relation $\sim$ on $\mathbb{C}$ is defined by $x \sim y \Leftrightarrow |x| = |y|$.
3. (10 pts) Draw the operation table of a group with 4 elements which is not isomorphic to $\mathbb{Z}_4$.
4. (10 pts) Draw the subgroup diagram of $\mathbb{Z}_{20}$.
5. (15 pts) Find the number of generators of $\mathbb{Z}_{55}$.
6. (15 pts) Find all abelian groups of order 216, up to isomorphism.

7. Cosets and Lagrange's theorem

We recall the materials in Section 1 of Chapter 0.

DEFINITION 7.1. Let G be a group and fix $H \leq G$.

(a) We give a relation on G by

$$a \sim_L b \iff a^{-1}b \in H.$$

A similar relation on G is defined by

$$a \sim_R b \iff ab^{-1} \in H.$$

Then $\sim_L$ and $\sim_R$ are equivalence relations, i.e., each relation is reflexive, symmetric and transitive. Note that an equivalence relation on G induces a partition on G.

(b) Let $a \in G$. The equivalence class of a with respect to $\sim_L$, i.e.,

$$[a]_L = \{x \in G \mid a \sim_L x\} = \{x \in G \mid a^{-1}x \in H\} = \{ah \mid h \in H\} =: aH$$

is called the *left coset* of H containing a. The set

$$[a]_R = \{x \in G \mid x \sim_R a\} = \{x \in G \mid xa^{-1} \in H\} = \{ha \mid h \in H\} =: Ha$$

is called the *right coset* of H containing a.

REMARK 7.2. (1) Note that $aH \neq Ha$, in general, since

$$b \in aH \Leftrightarrow a^{-1}b \in H \not\Leftrightarrow ba^{-1} \in H \Leftrightarrow b \in Ha.$$

(2) If G is abelian, we have $aH = Ha$ for each $H \leq G$ and $a \in G$.

(3) For $H \leq G$, the collection $\{[a]_L \mid a \in G\}$ is a partition of G, which is called the (left) coset decompositioncoset decomposition of H in G. It's the same for the right case.

EXAMPLE 7.3. (1) Consider a subgroup $n\mathbb{Z} = \{0, \pm n, \pm 2n, \ldots\}$ of the infinite cyclic group $\mathbb{Z}$ with the addition $+$. The equivalence relation $\sim_L$ is defined by $k \sim_L l \Leftrightarrow -k + l \in n\mathbb{Z}$. For each $k \in \mathbb{Z}$, we divide k by n to have $k = qn + r$ where $0 \leq r < n$, i.e., $k - r = qn \in n\mathbb{Z}$ or $k \in r + n\mathbb{Z}$ for a unique r such that $0 \leq r < n$. So we enumerate all cosets of $n\mathbb{Z}$ in $\mathbb{Z}$:

$$0 + n\mathbb{Z}, 1 + n\mathbb{Z}, 2 + n\mathbb{Z}, \ldots, (n-1) + n\mathbb{Z}.$$

We denote the coset $a + n\mathbb{Z}$ by simply $\overline{a}$:

$$\mathbb{Z}_n := \{\overline{0}, \overline{1}, \overline{2}, \ldots, \overline{n-1}\}.$$

Then $\mathbb{Z}_n$ is a group endowed with $\overline{a} + \overline{b} = \overline{a+b}$.

(2) Let $H = \{1, (1,2)\}$ be a subgroup of S_3. Then $(2,3)H = \{(2,3), (1,2,3)\} = (1,2,3)H$ and $(1,3)H = \{(1,3), (1,3,2)\} = (1,3,2)H$. Note that we use the *right* notation.[38] Now we have a complete set C of left cosets of H in S_3.

However, even if we define $aH \cdot bH = (ab)H$, C does not form a group because $(1,3)H \cdot (2,3)H = (1,2,3)H = (2,3)H$ and $(1,3)H \neq H$.

Later, we will study the notion of a *normal* subgroup, which is exactly the case when the set of cosets forms a group.

Exercise 7.4. (1) Let $H \leq G$ be a subgroup such that $g^{-1}hg \in H$ for all $g \in G$ and all $h \in H$. Show that every left coset gH is the same as the right coset Hg. Also show the converse statement.

(2) Let

$$D_{2\cdot 4} = \{1, (1,2,3,4), (1,3)(2,4), (1,4,3,2), (1,3), (2,4), (1,2)(3,4), (1,4)(2,3)\}$$

be the 4th dihedral group and $H = \{1, (1,3)(2,4)\}$. Find all left cosets of H in $D_{2\cdot 4}$. Are these the same as the right cosets of H in $D_{2\cdot 4}$?

Proposition 7.5. *Let G be a group and $H \leq G$. Then any left(or right) coset of H has the same number of elements as H.*

Proof. Define a map $\phi : H \to aH$ by $h \mapsto ah$. This map is onto by the definition of aH, and this is 1-1 since $ah = ah'$ implies $h = h'$. We have the same proof for the right coset Ha. □

Since a collection of cosets is a partition of a group and each cell has the same cardinality, we obtain the theorem of Lagrange. Note that this theorem imposes just a necessary condition for the order of a subgroup.

Theorem 7.6 (Lagrange). *Let G be a finite group and $H \leq G$. Then $|H|$ is a divisor of $|G|$.*

---

[38] Don't be confused with the left notation in the book [F].

Corollary 7.7. *Let G be a finite group and $a \in G$. Then $|a| := |\langle a \rangle|$ is a divisor of $|G|$. In particular, if $|G|$ is a prime p, then any non-identity element is a generator of G, in other words, G is cyclic and is isomorphic to $\mathbb{Z}_p$.*

REMARK 7.8. The converse of Lagrange's theorem is false. For example, A_4 has order 12 but does not contain a subgroup of order 6.

EXERCISE 7.9. Let G be a group with $|G| = pq$ (p, q, distinct primes). Then show that every proper subgroup of G is cyclic.

DEFINITION 7.10. Let H be a subgroup of G. The number of left(or right) cosets of H in G is called the *index* of H in G, denoted by $(G : H)$. If G is finite, then $|G| = |H|(G : H)$.

EXERCISE 7.11. (1) Find all cosets of the subgroup $\langle 4 \rangle$ in the group $\mathbb{Z}_{12}$ and the index $\langle 4 \rangle$ in $\mathbb{Z}_{12}$. Find the index of $\langle 3 \rangle$ in the group $\mathbb{Z}_{24}$.

(2) Let $\sigma = (1, 2, 5, 4)(2, 3)$ in S_5. Find the index of $\langle \sigma \rangle$ in S_5.

(3) Show that the number of left cosets of a subgroup H of G is equal to the number of right cosets of H of G, that is, exhibit a 1-1 map of the collection of left cosets onto the collection of right cosets.

Theorem 7.12. *Let $K \leq H \leq G$ be subgroups of G. Then*

$$(G : K) = (G : H)(H : K).$$

Proof. Let $\{g_1H, g_2H, \ldots, g_mH\}$ be the set of all distinct cosets of H in G, and $\{h_1K, \ldots, h_nK\}$ be the set of all distinct cosets of K in H. Then we claim that $\{(g_ih_j)K |\ 1 \leq i \leq m, 1 \leq j \leq n\}$ is the set of all distinct cosets of K in G. First,

$$G = \bigcup_{i=1}^{m} g_iH = \bigcup_{i=1}^{m} g_i \left(\bigcup_{j=1}^{n} h_jK \right) = \bigcup_{i=1}^{m} \bigcup_{j=1}^{n} g_ih_jK.$$

Now it is left to prove the disjointness. If $g_ah_bK = g_rh_sK$, then g_ah_b and g_rh_s are in the same coset of K in $G \Leftrightarrow g_ah_b \sim_L g_rh_s \Leftrightarrow g_ah_bk = g_rh_s$ for some $k \in K$. Since $h_bk \in H$, $g_aH = g_rH$, i.e., $a = r$, so $b = s$. □

EXERCISE 7.13. * Let G be a group and let H and K be subgroups with $(G : H)$ and $(G : K)$ finite.

(1) Show that $(H : H \cap K) \leq (G : K)$.

(2) Show that $(G : H \cap K) \leq (G : H)(G : K)$ and the equality holds if and only if $G = HK$ where $HK = \{hk|\ h \in H, k \in K\}$.

In the next section, we give an application of cosets in a finite vector space, which is basic in coding theory.

Midterm Exam

(October 17, 2013)

1. (30 points) Determine whether each statement is true or not: True(T) or False(F).
(If you answer correctly, you get 3 points; If nothing, 1 point; If wrong, 0 point.)
   1) The set $\mathbb{Z}$ of integers with the addition is isomorphic to the set $\mathbb{Q}$ of rational numbers with the addition.
   2) Every group of order 4 is isomorphic each other.
   3) Every cyclic group is an abelian group.
   4) Every subgroup of a cyclic group is also cyclic.
   5) $\mathbb{Z}_3 \times \mathbb{Z}_3$ is isomorphic to $\mathbb{Z}_9$.
   6) S_3 is an abelian group.
   7) Any permutation in S_n is written as a product of disjoint cycles.
   8) The number of odd permutations in S_n is $\frac{n!}{2}$
   9) For a subgroup H of G and $a, b \in G$, $aH \cap bH = \emptyset$ if $aH \neq bH$.
   10) For a subgroup H of G and $a \in G$, the left coset aH is the same as the right coset Ha.
2. (10pts) (1) Find all subgroups of $\mathbb{Z}_{12}$.
(2) What is the order of the cyclic subgroup generated by $\sigma = (1, 2, 3, 4)(5, 6, 7)(8, 9) \in S_9$?
3. (15pts) (1) Find $p(5)$, the partition number of 5.
(2) Classify all abelian groups of order 3^5 up to isomorphism.
4. (10pts) Consider the dihedral group

$$D = \{1, (1, 2, 3, 4), (1, 3)(2, 4), (1, 4, 3, 2), (1, 3), (2, 4), (1, 2)(3, 4), (1, 4)(2, 3)\}$$

of a square and its subgroup $H = \{1, (1, 3)(2, 4)\}$. Find all left cosets of H in D.
5. (10pts) State the Lagrange theorem.

• For each of the following problems, you should justify your answer.

6. (10pts) Let $\mathbb{C}$ be the set of complex numbers and $H = \left\{ \begin{pmatrix} a & b \\ -b & a \end{pmatrix} \mid a, b \in \mathbb{R} \right\}$. We consider the multiplication on $\mathbb{C}$ and the matrix multiplication on H. Show that $\mathbb{C}$ with multiplication is isomorphic to H with matrix multiplication.
7. (15pts) Consider a combination of numbers in a square with one blank. By moving the numbers using the blank, is it possible to change from the left side to the right with finally the same position of the blank? Why?

1	4	7
2	5	8
3	6	

$\longrightarrow$

1	2	3
4	5	6
7	8	

Midterm Exam

(October 19, 2009)

1. Determine whether each statement is true or not: True(T) or False(F).
   (If you answer correctly, you get 3 points; If nothing, 1 point; If wrong, 0 point.)
   1) In the plane $\mathbb{R}^2$, there are infinitely many types of the strip patterns (or the frieze groups), up to isomorphism.
   2) The set $\mathbb{N} = \{0, 1, 2, 3, 4, \ldots\}$ of natural numbers is a group under addition.
   3) The groups $\mathbb{Z}_{12}$ and $\mathbb{Z}_4 \times \mathbb{Z}_3$ under addition are isomorphic.
   4) A group G is abelian if every proper subgroup is abelian.
   5) The set of all $n \times n$ matrices with determinant -1 under multiplication is a group.
   6) The set of all $n \times n$ real matrices A such that $A^T A = I_n$ is a subgroup of $GL_n(\mathbb{R})$.
   7) An infinite cyclic group has 2 generators.
   8) The symmetric group S_n has order $n!$.
   9) The number of even permutations in S_n is greater than the number of odd permutations in S_n.
   10) The cyclic group $\mathbb{Z}_{10}$ has a subgroup of order 3.
2. (10 pts) Let G be a group and let g be a fixed element of G. The map i_g is defined by $i_g(x) = gxg^{-1}$ for $x \in G$. Show that i_g is an isomorphism of G with itself. [Hint: Show that i_g is 1-1 and onto, and more.]
3. (10 pts) Find all subgroups of the cyclic group $\mathbb{Z}_{12}$.
4. (10 pts) Let $\sigma = (1, 4, 5)(7, 8)(2, 5, 7)(3, 6)$ be a product of cycles in S_8. (Regard σ as a product written in the right notation.)
   (1) Express σ as a product of disjoint cycles.
   (2) What is the order of the subgroup $\langle \sigma \rangle$ of S_8.
5. (15 pts) (1) Find all cosets of the subgroup $\langle 4 \rangle$ in the group $\mathbb{Z}_{24}$, and the index $(\mathbb{Z}_{24} : \langle 4 \rangle)$.
   (2) Let $D_{2 \cdot 4} = \{1, (1, 2, 3, 4), (1, 3)(2, 4), (1, 4, 3, 2), (1, 3), (2, 4), (1, 2)(3, 4), (1, 4)(2, 3)\}$ be the 4th dihedral group and $H = \{1, (1, 3)(2, 4)\}$. Find all left cosets of H in $D_{2 \cdot 4}$.
6. (10 pts) Find every element in $\mathbb{Z}_2 \times \mathbb{Z}_2 \times \mathbb{Z}_4$ whose order is 2.
7. (15 pts) Classify all finite abelian groups of order 720, up to isomorphism.

8. Syndrome decoding

An error occurs during data transmission[39] through a noisy channel. In order to receive a right message, we require a process of error detection, moreover, error correction.

original message	$\xrightarrow{\text{encode}}$	code vector to be transmitted	$\overset{\textit{noisy channel}}{\rightsquigarrow}$	received code vector	$\xrightarrow{\text{decode}}$	decoded message

REMARK 8.1. * A vector space V is an abelian group under the addition. For a subspace W of V, we have its disjoint cosets and the corresponding representatives. The coset decomposition in a finite vector space is used for maximum-likelihood decoding when we perform an error-correction of a distorted vector. This general decoding method by finding the syndrome of a coset leader is called *syndrome decoding.*

EXAMPLE 8.2. As an easy example, consider the vector space $\mathbb{F}_2^4$ where $\mathbb{F}_2 = \{0, 1\}$ is the field with two elements. Let $G = \begin{pmatrix} 1 & 0 \\ 0 & 1 \\ 1 & 1 \\ 0 & 1 \end{pmatrix}$ and $C = \operatorname{col}(G) \leq \mathbb{F}_2^4$, the column space of G. A subspace of a finite vector space is called a *code.* Note that a *generator matrix* G for a code C represents the encoding map $G : \mathbb{F}_2^2 \to \mathbb{F}_2^4$ by $\mathbf{x} \mapsto G\mathbf{x} =: \mathbf{y} \in C$, and that the components y_i $(i = 1, 2, 3, 4)$ of $\mathbf{y}$ satisfies:

$$y_3 = y_1 + y_2, \quad y_4 = y_2.$$

Thus, we take a *parity check matrix* $P = \begin{pmatrix} 1 & 1 & 1 & 0 \\ 0 & 1 & 0 & 1 \end{pmatrix}$ such that

$$\operatorname{row}(P) = C^{\perp} \leq \mathbb{F}_2^4,$$

that is, $PG = O$. The elements in C are codewords but the others are non-codewords, meaning that they are distorted through a noisy channel. Note that

$$P\mathbf{y} = 0 \text{ if and only if } \mathbf{y} \in C.$$

We define the *syndrome* of $\mathbf{y} \in \mathbb{F}_2^4$ to be

$$\operatorname{syn}(\mathbf{y}) = P\mathbf{y}.$$

---

[39]Its mathematical development started from a landmark paper in information theory: Claude Shannon, *A Mathematical Theory of Communication*, The Bell System Technical Journal **27** (1948), 379–423, 623–656.

Our question is: How can we correct a vector into a codeword if it has an error?

Now we obtain the cosets of C in $\mathbb{F}_2^4$ and take their representatives with minimal number of nonzero components. Those representatives are called *coset leaders.*

$$\begin{aligned} C &= \{0000, 1010, 0111, 1101\}, \\ 1000 + C &= \{1000, 0010, 1111, 0101\}, \\ 0100 + C &= \{0100, 1110, 0011, 1001\}, \\ 0001 + C &= \{0001, 1011, 0110, 1100\}. \end{aligned}$$

In fact, we don't have to enumerate all elements in the cosets of C for error-correction. Note that $\mathbf{a}, \mathbf{b} \in \mathbf{a} + C$ if and only if $P\mathbf{a} = P\mathbf{b}$. This is because $P(\mathbf{a} + \mathbf{c}) = P\mathbf{a}$ for $\mathbf{c} \in C$. The number of syndromes is 2^d with $d = \dim C^{\perp}$.

Thus, from a received vector $\mathbf{y}' \in \mathbb{F}_2^n$, we can find a corrected codeword $y \in C$ as follows:

1) Calculate the syndrome of $\mathbf{y}'$, $\operatorname{syn}(\mathbf{y}')$.
2) If $\operatorname{syn}(\mathbf{y}') = 0$, then y' is a correct codeword.
3) If $\operatorname{syn}(\mathbf{y}') \neq 0$, then find the coset leaders whose syndromes are distinct and

take the coset leader $\mathbf{e}$ such that $\operatorname{syn}(\mathbf{y}') = P\mathbf{e}$.

4) A distorted $\mathbf{y}' \notin C$ is corrected into the codeword $\mathbf{y}' + \mathbf{e} \in C$.

Remark 8.3. Coding theory attracted a remarkable attention in 1971 when the space probe Mariner 9 entered an orbit of Mars and transmitted photos, which were clearly error-corrected.[40] The theory of error-detecting and error-correcting codes is indispensable to modern computer civilization.

[40]Reed-Müller (32,6)-block codes was used in 1971. The space probe Voyager used convolution codes.

Midterm Exam

(October 15, 2012)

1. (40 points) Determine whether each statement is true or not: True(T) or False(F). (If you answer correctly, you get 3 points; If nothing, 1 point; If wrong, 0 point.)
   1) The set $\mathbb{Z}$ of integers is a group under multiplication.
   2) Every cyclic group is an abelian group.
   3) Any subgroup of a cyclic group is cyclic.
   4) The group $\mathbb{Z}_2 \times \mathbb{Z}_3$ is isomorphic to $\mathbb{Z}_6$.
   5) The dihedral group D_{2n} is abelian.
   6) The order of D of symmetries of a square is equal to that of the symmetric group S_4.
   7) Every permutation in S_n is a product of adjacent transpositions.
   8) The set of odd permutations in S_n is a group of order $\frac{n!}{2}$.
   9) For a subgroup H of G, the left coset aH of H is the same as the right coset Ha.
   10) The set of cosets of a subgroup H of G is a partition of G.

   - Let $G \in M_{n\times k}(\mathbb{F}_2)$ with $\text{rank}(G) = k$ and $P \in M_{(n-k)\times n}(\mathbb{F}_2)$ with $\text{rank}(P) = n - k$, satisfying $PG = O$. We let $C = \text{col}(G) \leq \mathbb{F}_2^n$.

   11) Then $Py = 0$ if and only if $y \in C$.
   12) If $a, b \in a + C$, then $Pa = Pb$.
   13) If $a + C$ and $b + C$ are disjoint, then $Pa \neq Pb$.

2. (25pts) (1) What is the number of transformations on a regular triangle?
   (2) What is the number of generators of the cyclic group $(\mathbb{Z}_9, +)$?
   (3) Write all elements of $(\mathbb{Z}_9^\times, \times)$ and find its generator.
   (4) Write down the theorem of Lagrange.
   (5) Write $\sigma = \begin{pmatrix} 1 & 2 & 3 & 4 & 5 & 6 & 7 & 8 \\ 6 & 7 & 1 & 5 & 8 & 3 & 2 & 4 \end{pmatrix} \in S_8$ into a product of disjoint cycles.
3. (10pts) Let $\mathbb{C} = \{a + bi|\ a, b \in \mathbb{R}\}$, and $S = \left\{ \begin{pmatrix} a & b \\ -b & a \end{pmatrix} |\ a, b \in \mathbb{R} \right\}$. Show that $\mathbb{C}$ with the multiplication is isomorphic to S with the matrix multiplication.
4. (10pts) Find all abelian groups of order $32 = 2^5$, which are non-isomorphic each other.
5. (15pts) (1) Find the subgroup diagram of $\mathbb{Z}_{12}$.
   (2) Find all cosets of $\langle 3 \rangle$ in $\mathbb{Z}_{12}$.

CHAPTER 2

Quotient groups

In this chapter, we study the notion of a normal subgroup, which is one of the most important in a basic course of group theory. The structure of a group G is reflected in the structure of the quotient groups and the subgroups of G. We shall see that the study of the quotient groups of G is essentially equivalent to the study of the homomorphisms of G.

1. Homomorphisms

We can get some information on algebraic structures via a structure-relating map between them.

DEFINITION 1.1. Let $(G, *)$ and $(G', *')$ be groups. A map $\phi : G \to G'$ is called a *homomorphism* of G into G' if it is an operation-preserving map[1], that is,

$$\phi(a * b) = \phi(a) *' \phi(b) \quad \text{for all } a, b \in G.$$

REMARK 1.2. (1) A homomorphism needs not to be 1-1 or onto. A bijective homomorphism is called an *isomorphism*. We say that two groups are em isomorphic if there exists an isomorphism between them.

(2) We have the *trivial homomorphism* $\phi : G \to G'$, defined by $\phi(g) = e'$ for all $g \in G$, where e' is the identity in G'.

(3) The composition of group homomorphisms is again a homomorphism.

(4) Let G, G' be groups and $\phi : G \to G'$ be a homomorphism. Then

$$\phi(e) = e'$$

[1]Understanding the *duality* between a space(e.g. a set with an operation) and a function(e.g. an operation-preserving map) is one of the main themes in mathematics. Figuratively speaking, we want to understand the relationship between a land and a tree. If we see a land then we see which tree is grown on that, and conversely, if we observe a tree then we come to know the property of the land.

and

$$\phi(a^{-1}) = \phi(a)^{-1} \text{ for each } a \in G.$$

(5) Let $\phi : G \twoheadrightarrow G'$ be a group homomorphism of G onto G'. If G is abelian then G' is abelian, since for each $a', b' \in G'$, there are $a, b \in G$ such that $\phi(a) = a'$, $\phi(b) = b'$, and $a'b' = \phi(a)\phi(b) = \phi(ab) = \phi(ba) = \phi(b)\phi(a) = b'a'$.

EXAMPLE 1.3. (1) Let F be the additive group of all real-valued functions on $\mathbb{R}$, and let c be any real number. Define $\phi_c : F \to \mathbb{R}$ to be $\phi_c(f) = f(c)$ for $f \in F$. This is a homomorphism since $\phi_c(f+g) = (f+g)(c) = f(c)+g(c) = \phi_c(f)+\phi_c(g)$, which we call the *evaluation homomorphism.*

(2) Let $G = G_1 \times \cdots \times G_n$ be the direct product of groups. The *projection map* $\pi_i : G \to G_i$ defined by $\pi_i(g_1, \ldots, g_n) = g_i$ is a homomorphism for each $i = 1, \ldots, n$.

(3) Let C be the additive group of all continuous real-valued functions on $\mathbb{R}$ and D be the subgroup of differentiable functions. Then the differential operator $\frac{d}{dx}$: $D \to D$ by $f(x) \mapsto f'(x)$ is a homomorphism, and an integral operator $I_x : C \to C$ define by $I(f) = \int_0^x f(t)dt$ is also a homomorphism.

(4) The map exp : $(\mathbb{R}, +) \to (\mathbb{R}_+, \cdot)$ by $x \mapsto e^x$ is a group isomorphism, whose inverse map is log.

EXERCISE 1.4. Show that $\phi : G \to G$ by $g \mapsto g^{-1}$ is a group homomorphism if and only if G is an abelian group.

REMARK 1.5. (1) Let G be generated by $\{a_i \in G | \ i \in I\}$ for an index set I. If $\phi : G \to G'$ and $\mu : G \to G'$ are two homomorphisms such that $\phi(a_i) = \mu(a_i)$ for every $i \in I$, then $\phi = \mu$. In particular, a homomorphism ϕ of a cyclic group $G = \langle g \rangle$ is determined by its value $\phi(g)$ on a generator g of the group G.

(2) Note that the homomorphisms ϕ from a cyclic group $G = \langle g \rangle$ are determined by $\phi(g)$, and moreover $|\phi(g)|$ must be a divisor of $|g| = |G|$ if $|G|$ is finite.

EXERCISE 1.6. For the following given groups, find all nontrivial homomorphisms ϕ.

1) $\phi : \mathbb{Z}_{12} \to \mathbb{Z}_5$, 2) $\phi : \mathbb{Z}_{12} \to \mathbb{Z}_9$, 3) $\phi : \mathbb{Z}_{12} \to \mathbb{Z}_9^{\times}$,
4) $\phi : \mathbb{Z}_{12} \to S_3$, 5) $\phi : \mathbb{Z}_4 \to S_3$.

EXERCISE 1.7. * Let $\mathbb{F}$ be a field[2]. Usually, we take $\mathbb{F} = \mathbb{Q}, \mathbb{R}$ or $\mathbb{C}$. A group homomorphism

$$\phi : G \longrightarrow GL_n(\mathbb{F})$$

is called a (linear or matrix) *representation* of G of degree n over $\mathbb{F}$.

(1) Note that $GL_2(\mathbb{R})$ is the group of invertible 2×2 real matrices. We define a map ϕ from $\mathbb{C}^\times$ into $GL_2(\mathbb{R})$ by

$$\phi(a + bi) = \begin{pmatrix} a & b \\ -b & a \end{pmatrix}.$$

Show that ϕ is a group homomorphism and 1-1. So we can embed $\mathbb{C}^\times$ into $GL_2(\mathbb{R})$.

(2) Note that $GL_2(\mathbb{C})$ is the group of invertible 2×2 complex matrices. Let

$$Q = \{\pm 1, \pm i, \pm j, \pm k\}$$

be the *quaternion group* (i.e., $i^2 = j^2 = k^2 = -1,\ ij = k$), we define $\phi : Q \to GL_2(\mathbb{C})$ by

$$\phi(i) = \begin{pmatrix} i & 0 \\ 0 & -i \end{pmatrix}, \quad \phi(j) = \begin{pmatrix} 0 & 1 \\ -1 & 0 \end{pmatrix}, \quad \phi(k) = \begin{pmatrix} 0 & i \\ i & 0 \end{pmatrix},$$

which is a faithful(i.e., 1-1) representation of Q. Show that ϕ is a group homomorphism and 1-1.

(3) In general, we consider

$$\mathbb{H} = \{a + bi + cj + dk |\ a, b, c, d \in \mathbb{R}\} = \{\alpha + \beta j |\ \alpha, \beta \in \mathbb{C}\},$$

where $i^2 = j^2 = k^2 = ijk = -1$. [3] The set of nonzero elements of $\mathbb{H}$ is a multiplicative group, so we have a representation $\phi : \mathbb{H}^\times \to GL_2(\mathbb{C})$ by

$$\phi(\alpha + \beta j) = \begin{pmatrix} \alpha & \beta \\ -\bar{\beta} & \bar{\alpha} \end{pmatrix}.$$

Show that ϕ is a group homomorphism and 1-1.

EXERCISE 1.8. * Let $S^3 = \{x \in \mathbb{H} |\ |x| = 1\}$ be the unit sphere in $\mathbb{R}^4$, and

$$SU(2) = \{A \in GL_2(\mathbb{C}) |\ A\bar{A}^t = I,\ \det A = 1\}$$

be the *special unitary group* of degree 2. Show that S^3 and $SU(2)$ are isomorphic. Also show that S^1 and $SO(2)$ are isomorphic.

---

[2] See the footnote of Ch.1, Example 1.4.

[3] The set $\mathbb{H}$ with the usual addition and the multiplication by $i^2 = j^2 = k^2 = ijk = -1$ is called the *Hamiltonian skew-field*. Note that every nonzero element of $\mathbb{H}$ has the multiplicative inverse, but the multiplication is not commutative.

The inverse image of the trivial subgroup is critical to the study of homomorphisms $\phi : G \to G'$.

DEFINITION 1.9. Let $\phi : G \to G'$ be a group homomorphism. The subgroup

$$\phi^{-1}(e') = \{x \in G|\ \phi(x) = e'\}$$

is called the *kernel* of ϕ, denoted by $\ker \phi$.

EXAMPLE 1.10. (1) Let $\exp : \mathbb{R} \to \mathbb{R}_+$ be the exponential function. Then exp is a homomorphism from the additive group $\mathbb{R}$ into the multiplicative group $\mathbb{R}_+$ of positive real numbers, with $\ker(\exp) = \{0\}$.

(2) Let $\phi = |\ \ | : \mathbb{C}^\times \to \mathbb{R}_+$ be the absolute value function, which is a homomorphism since $|z_1 z_2| = |z_1||z_2|$. Note that $\ker \phi = \{z \in \mathbb{C}^\times|\ |z| = 1\}$ is a subgroup of $\mathbb{C}^\times$, and that the cosets of $\ker \phi$ are circles with center at the origin in the complex plane. Each circle is collapsed by the homomorphism $|\ \ |$ onto its intersection point with the positive real axis.

(3) The sign of an even permutation is $+1$ and the sign of an odd permutation is -1. Observe that the map

$$\mathrm{sgn}_n : S_n \to \{\pm 1\} \quad \text{defined by } \sigma \mapsto \text{ sign of } \sigma$$

is a homomorphism of S_n onto the multiplicative group $\{\pm 1\}$, whose kernel is A_n.

EXERCISE 1.11. Find every non-trivial homomorphism $\phi : \mathbb{Z}_{12} \to \mathbb{Z}_{18}$ and its kernel $\ker \phi$.

EXERCISE 1.12. Let $\phi : G \to G'$ be a group homomorphism. Show that $\phi(G)$ is abelian if and only if for all $x, y \in G$, we have $xyx^{-1}y^{-1} \in \ker \phi$.

The subgroup $\ker \phi$ of G has a special property as the following theorem shows. Thus, cosets of $H = \ker \phi$ collapsed by ϕ are well-defined, which we call *fibers* in G under a homomorphism $\phi : G \to G'$. That is, the fibers of ϕ are the sets of elements of G projecting to single elements of G'.

Theorem 1.13. *Let $\phi : G \to G'$ be a group homomorphism and $H = \ker \phi \le G$. Then $\phi^{-1}(\phi(a)) = \{x \in G|\ \phi(x) = \phi(a)\}$ is the left coset aH of H and is also the right coset Ha, that is,*

$$\phi^{-1}(\phi(a)) = aH = Ha.$$

Proof. To prove the equality, first we assume that $\phi(x) = \phi(a)$. Then $\phi(a)^{-1}\phi(x) = e'$. Since ϕ is a homomorphism, $\phi(a^{-1}x) = \phi(a^{-1})\phi(x) = e'$. So $a^{-1}x \in H = \ker\phi$, i.e., $x \in aH$. Conversely, let $x \in aH$, i.e., $x = ah$ for some $h \in H$. Then $\phi(x) = \phi(ah) = \phi(a)\phi(h) = \phi(a)e' = \phi(a)$, so that $x \in \phi^{-1}(\phi(a))$.

The proof that $\phi^{-1}(\phi(a)) = Ha$ follows the same procedure. □

EXERCISE 1.14. Prove that $\phi^{-1}(\phi(a)) = Ha$ in the previous theorem.

Corollary 1.15. *A group homomorphism $\phi : G \to G'$ is 1-1 if and only if* $\ker\phi = \{e\}$. *In this case, we call ϕ a* monomorphism *of G into G'.*[4]

Proof. The implication $\Rightarrow$ is clear, and the converse follows from $\phi^{-1}(\phi(a)) = a\{e\} = \{a\}$ for each $a \in G$. □

Theorem 1.16. *Let $\phi : G \to G'$ be a group homomorphism.*

(a) *If $H \leq G$, then $\phi(H) \leq G'$. That is, the image of a subgroup of G is a subgroup of G'.*
(b) *If $K' \leq G'$, then $\phi^{-1}(K') \leq G$. That is, the inverse image of a subgroup of G' is a subgroup of G.*

Proof. (a) Let H be a subgroup of G and $\phi(a), \phi(b)$ be any two elements in $\phi(H)$. Then $\phi(a)\phi(b) = \phi(ab) \in \phi(H)$, that is, $\phi(H)$ is closed under the operation of G'. Clearly, we have $e' = \phi(e) \in \phi(H)$ and $\phi(a)^{-1} = \phi(a^{-1}) \in \phi(H)$, thus we proved that $\phi(H) \leq G'$.

(b) Let $K' \leq G'$ and $a, b \in \phi^{-1}(K')$. Then $\phi(ab) = \phi(a)\phi(b) \in K'$, which means that $ab \in \phi^{-1}(K')$. Also, we have $e \in \phi^{-1}(K')$ from $\phi(e) = e' \in K'$, and $a^{-1} \in \phi^{-1}(K')$ from $\phi(a^{-1}) = \phi(a)^{-1} \in K'$. Thus we proved that $\phi^{-1}(K') \leq G$. □

Corollary 1.17. *Let $\phi : G \to G'$ be a group homomorphism.*

(a) *If $|G'|$ is finite, then $|\phi(G)|$ is finite and is a divisor of $|G'|$.*
(b) *If $|G|$ is finite, then $|\phi(G)|$ is finite and is a divisor of $|G|$.*
(c) *If $|G|$ is a prime, then ϕ is either the trivial homomorphism or a 1-1 map.*

Proof. (a) The image $\phi(G)$ is a subgroup of G'.

(b) $|\phi(G)| = |G|/|\ker\phi|$.

---

[4]A surjective homomorphism is called an *epimorphism*. We say that a homomorphism is an *endomorphism* if its domain is the same as the codomain. A bijective endomorphism is called an *automorphism*.

(c) This follows from (b). □

Now we define one of the most important notions in basic group theory.

DEFINITION 1.18. A subgroup H of G is called *normal*, denoted by $H \trianglelefteq G$, if

$$aH = Ha \quad \text{for all } a \in G.$$

REMARK 1.19. (1) Any subgroup of an abelian group is normal.

(2) For any group homomorphism $\phi : G \to G'$, $\ker \phi$ is a normal subgroup of G.

(3) Consider $H = \{1, (1,2)\} \leq S_3$. Then

$$(2,3)H = \{(2,3), (1,2,3)\} \neq \{(2,3), (1,3,2)\} = H(2,3),$$

so H is not a normal subgroup of S_3. Is there a nontrivial homomorphism $\phi : S_3 \to \mathbb{Z}_3$?

(4) If $H \leq G$ with $(G : H) = 2$, then $H \trianglelefteq G$. For example, $A_n \trianglelefteq S_n$.

EXERCISE 1.20. (1) Enumerate all elements of the alternating group A_4, and show that

$$A_4 = \langle (12)(34), (123) \rangle.$$

(2) What is a nontrivial homomorphism $\phi : A_4 \to \mathbb{Z}_3$? What is $\ker \phi$?

(3) Is there a nontrivial homomorphism $\phi : A_4 \to \mathbb{Z}_2$?

(4) Find all subgroups of A_4.

EXERCISE 1.21. ([CDU])* There is an interesting homomorphism between $SL_2(\mathbb{C})$ and the Lorentz[5] group L, which is the group of all linear transformations of the vector space $\mathbb{R}^4$ that preserve the Lorentz metric

$$|x| := x_0^2 - x_1^2 - x_2^2 - x_3^2.$$

To $x \in \mathbb{R}^4$ we associate a 2×2 matrix $\psi(x)$ as follows:

$$\psi(x) = \begin{pmatrix} x_0 + x_3 & x_1 - ix_2 \\ x_1 - ix_2 & x_0 - x_3 \end{pmatrix},$$

so that $|x| = \det(\psi(x))$. Now, the map $\varphi : SL_2(\mathbb{C}) \to L$ given by $\varphi(A)(x) = \psi^{-1}(A\psi(x)A^*)$ is a group homomorphism.

[5]Hendrik Antoon Lorentz(1853–1928), a Dutch physicist. He shared the 1902 Nobel Prize in Physics with Pieter Zeeman for the discovery and theoretical explanation of the Zeeman effect.

2. Normal subgroups

Consider a homomorphism $\phi : G \to G'$ with $\ker\phi = H \leq G$. Then we have a 1-1 correspondence between

$$\{\phi^{-1}(\phi(a))|\ a \in G\} = \{aH = Ha|\ a \in G\} \overset{1-1}{\longleftrightarrow} \phi(G) = \{\phi(a)|\ a \in G\} \leq G'.$$

Since $\phi(G)$ is a subgroup of G', the group operation on $\phi(G)$ is induced from that on G'. Now we let $x = \phi(a)$, $y = \phi(b) \in \phi(G)$ for some $a, b \in G$, so $xy = \phi(ab)$. We give an operation on the set G/H of cosets of H by

$$(aH)(bH) := (ab)H.$$

Note that this operation may have an ambiguity because the coset aH (which is an equivalence class) has many representatives in aH. But, this operation is *well-defined* since, for any $h_1, h_2 \in H$, we have $(ah_1bh_2)H = (abh_3h_2)H = (ab)H$ where $h_3 \in H$. Therefore we obtain that $G/H \cong \phi(G)$ as a group. This is called the *fundamental homomorphism theorem.*

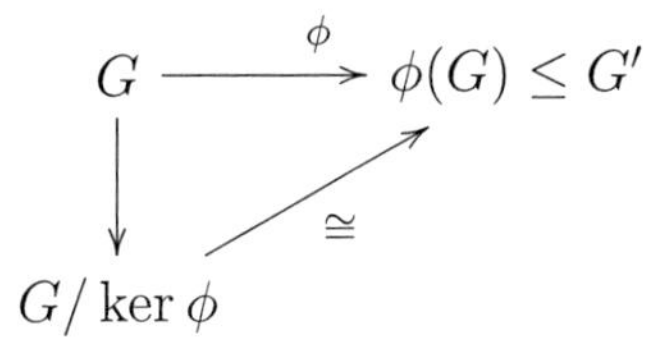

DEFINITION 2.1. Let $\phi : G \to G'$ be a group homomorphism with $\ker\phi = H \leq G$. We denote the set of all cosets of H by G/H, which is a group under the well-defined operation as above, with identity H and $(aH)^{-1} = a^{-1}H$. We call G/H the *quotient group* (or the *factor group*) of G modulo H. Two elements $x, y \in G$ are said to be congruent modulo H if $x, y \in aH = Ha$ for some $a \in G$.

EXAMPLE 2.2. (1) Let $\gamma : \mathbb{Z} \to \mathbb{Z}_n$ be an additive homomorphism defined by $m \mapsto \overline{r}$ where $m = nq + r$ for $0 \leq r < n$. Then $\ker\gamma = n\mathbb{Z}$, so

$$\mathbb{Z}/n\mathbb{Z} \cong \mathbb{Z}_n \ \text{ by } r + n\mathbb{Z} \leftrightarrow \overline{r}.$$

Note that each element in $\mathbb{Z}/n\mathbb{Z}$ is a residue class modulo n. For example,

$$(2 + 5\mathbb{Z}) + (4 + 5\mathbb{Z}) = 1 + 5\mathbb{Z} \in \mathbb{Z}/5\mathbb{Z}.$$

(2) Consider a subgroup $H = \{1, (1,2,3), (1,3,2)\}$ of S_3. Then

$$(1,2)H = \{(1,2), (1,3), (2,3)\},$$

so the set of left cosets of H is $\{H, (1,2)H\}$. Note that $H \cdot H = H$, $H \cdot (1,2)H = (1,2)H$, and $(1,2)H \cdot (1,2)H = H$, regardless of any representative of a left coset. In fact, H is the kernel of sign : $S_3 \to \{\pm 1\}$.

(3) Let $H = \{1, (1,2)\} = (1,2)H$ be a subgroup of S_3. Then

$(2,3)H = \{(2,3), (1,2,3)\} = (1,2,3)H$ and $(1,3)H = \{(1,3), (1,3,2)\} = (1,3,2)H$.

Give an operation on $\{H, (2,3)H, (1,3)H\}$ by the operation of representatives. Since

$$(2,3)H(1,3)H = (2,3)(1,3)H = (1,3,2)H = (1,3)H \quad \text{but}$$

$$(1,2,3)H(1,3)H = (1,2,3)(1,3)H = (1,2)H = H,$$

this operation is not well-defined. In fact, H is not a kernel of any homomorphism $\phi : S_3 \to \mathbb{Z}_3$, in other words, there is no nontrivial homomorphism $\phi : S_3 \to \mathbb{Z}_3$.

EXERCISE 2.3. In the 4th dihedral group D, consider subgroups

$$N = \{1, (1,3)(2,4)\} \text{ and } H = \{1, (1,2)(3,4)\}.$$

(1) Find all left cosets of N in D and determine whether the operation on D/N defined by $(aN)(bN) = (ab)N$ is well-defined or not.

(2) Determine also for D/H.

Our goal is understanding the condition of a subgroup H of G, for which the operation on $G/H = \{aH|\ a \in G\}$ defined by $(aH)(bH) = (ab)H$ is well-defined. In other words, we characterize the subgroups of G by which the quotient groups are defined.

Theorem 2.4. *Let $H \leq G$. We give an operation on the set G/H of left cosets of H by*

$$(aH)(bH) := (ab)H.$$

This operation is well-defined if and only if H is a normal subgroup of G, i.e., $aH = Ha$ for all $a \in G$.

Proof. $\Leftarrow$) Choosing $a \in aH$ and $b \in bH$ produces that $(aH)(bH) = (ab)H$. Now we choose different representatives, that is, let $ah_1 \in aH$ and $bh_2 \in bH$. Then $(aH)(bH) = (ah_1bh_2)H$. Since $h_1b \in Hb = bH$, we write $h_1b = bh_3$ for some $h_3 \in H$. Hence $(ah_1bh_2)H = (abh_3h_2) = (ab)H$.

$\Rightarrow$) Let $x \in aH = xH$. Since $(xH)(a^{-1}H) = (xa^{-1})H$ and $(aH)(a^{-1}H) = (aa^{-1})H = H$, we obtain that $xa^{-1} = h \in H$ for some $h \in H$, thus $x = ha \in Ha$. We showed that $aH \subseteq Ha$ for any $a \in G$.

For the converse inclusion, we let $x \in Ha$. Then $xa^{-1} \in H$ and $a^{-1} \in x^{-1}H = a^{-1}H$. Since $(x^{-1}H)(aH) = (x^{-1}a)H$ and $(a^{-1}H)(aH) = H$, we have $x^{-1}a = h \in H$ for some $h \in H$, so $x = ah^{-1} \in aH$. □

Corollary 2.5. *Let* $H \trianglelefteq G$. *Then* $G/H = \{aH|\ a \in G\}$ *is a group under* $(aH)(bH) = (ab)H$, *which is called the* quotient group *(or the* factor group*) of* G *by* H.

Proof. Associativity holds in G/H since the operation in G is associative. The identity in G/H is H and $(aH)^{-1} = (a^{-1})H$ for each $a \in G$. □

Let us summarize: Let $H \leq G$ be a subgroup. Then

$$\begin{aligned} & H \text{ is the kernel of a homomorphism from } G \\ \Longleftrightarrow\ & H \text{ is a normal subgroup of } G \\ \Longleftrightarrow\ & \text{The operation } (aH)(bH) = (ab)H \text{ on } G/H \text{ is well-defined.} \end{aligned}$$

Exercise 2.6. (1) For $N = \{1, (1,2,3), (1,3,2)\} \trianglelefteq S_3$, show that $S_3/N \cong \mathbb{Z}_2$.

(2) For $N = \{1, (1,3)(2,4)\} \trianglelefteq D_{2\cdot 4}$, show that $D_{2\cdot 4}/N \cong \mathbb{Z}_2 \times \mathbb{Z}_2$.

(3) For $N = \{1, -1\} \trianglelefteq Q = \{\pm 1, \pm i, \pm j, \pm k\}$ with $i^2 = j^2 = k^2 = ijk = -1$, show that $Q/N \cong \mathbb{Z}_2 \times \mathbb{Z}_2$.

We study more on normal subgroups.

Definition 2.7. A homomorphism ϕ is called an automorphism of G if ϕ is an isomorphism of G onto itself. For each $g \in G$, define an automorphism

$$i_g : G \to G \text{ by } x \mapsto gxg^{-1},$$

which is called an *inner automorphism* of G. Performing i_g on x is said to be the *conjugation* of x by g.

Proposition 2.8. *Let* $H \leq G$. *Then the following are equivalent:*

(a) $gH = Hg$ *for all* $g \in G$.
(b) $gHg^{-1} = H$ *for all* $g \in G$.
(c) $ghg^{-1} \in H$ *for all* $g \in G$ *and* $h \in H$.
(d) $i_g(H) = H$ *for all* $g \in G$, *that is,* H *is invariant under all inner automorphisms of* G.

Proof. (a)⇒(c), (c)⇒(b), (b)⇒(a). Also, (d) is equivalent to the others. □

EXAMPLE 2.9. (1) Let us compute $i_{(1,2,3)}(H)$ for $H = \{1, (1,2)\} \leq S_3$.

$$i_{(1,2,3)}(H) = (1,2,3)H(1,3,2) = \{1, (1,3)\} \neq H,$$

which is the same as $(1,2,3)H \neq H(1,2,3)$, that is, H is not a normal subgroup of S_3.

(2) Let $H = \{1, (1,2,3), (1,3,2)\} \leq S_3$. Then $i_g(H) = H$ for any $g \in S_3$, thus $H \trianglelefteq S_3$. In fact, H is a unique subgroup of order 3.

(3) In general, the alternating group A_n consisting of all even permutations in S_n satisfies that $gA_ng^{-1} = A_n$ for any $g \in S_n$, so $A_n \trianglelefteq S_n$. We shall see that A_n is a unique (nontrivial proper) normal subgroup of S_n for $n \geq 5$.

EXERCISE 2.10. (1) A subgroup $H \leq G$ is called conjugate to K if $i_g(H) = K$ for some $g \in G$. Show that the conjugacy is an equivalence relation on the collection of subgroups of G.

(2) Show that if a group has exactly one subgroup H of a given order, then H is a normal subgroup of G.

(3) Let G be a group containing at least one subgroup of a fixed finite order s. Show that the intersection of all subgroups of G of order s is a normal subgroup of G. [Hint : Use the fact that if H has order s, then so does $x^{-1}Hx$ for all $x \in G$.]

Theorem 2.11 (Isomorphism theorems).

(a) (The fundamental homomorphism theorem) *Let $\phi : G \to G'$ be a group homomorphism. Then*

$$G/\ker\phi \cong \phi(G).$$

(b) * *Let M and N be subgroups of G with $N \trianglelefteq G$. Then*

$$M/(M \cap N) \cong (M + N)/N.$$

(c) * *Let M and N be normal subgroups of G such that $M \leq N$. Then $N/M \leq G/M$ and*

$$(G/M)/(N/M) \cong G/N.$$

REMARK 2.12. In the fundamental homomorphism theorem, let $H = \ker\phi \trianglelefteq G$. Then the map $\gamma : G \to G/H$ given by $x \mapsto xH$ is a homomorphism with $\ker\gamma = H$. This map is called the *canonical projection.* Note that there is an isomorphism $\mu : G/H \to \phi(G)$ such that $\phi = \mu \circ \gamma$.

EXAMPLE 2.13. (1) Fix $x \in G$. Define a homomorphism $\gamma_x : \mathbb{Z} \to G$ by $\gamma_x(a) = x^a$. Since $\ker(\gamma_x) \trianglelefteq \mathbb{Z}$, $\ker(\gamma_x) = n\mathbb{Z}$ for some $n \in \mathbb{Z}_{\geq 0}$. And $\mathrm{Im}(\gamma_x) = \langle x \rangle$. So we have

$$\mathbb{Z}/n\mathbb{Z} \cong \langle x \rangle \leq G$$

by the fundamental homomorphism theorem.

(2) Consider a homomorphism $\det : GL_n(\mathbb{R}) \to \mathbb{R}^\times$. Then

$$\ker(\det) = \{A \in GL_n(\mathbb{R}) |\ \det A = 1\} =: SL_n(\mathbb{R}),$$

called the *special linear group*. Since det is surjective, we have

$$GL_n(\mathbb{R})/SL_n(\mathbb{R}) \cong \mathbb{R}^\times$$

by the fundamental homomorphism theorem.

(3) Consider an additive homomorphism $\mathrm{tr} : M_{n\times n}(\mathbb{R}) \to \mathbb{R}$. Then

$$\ker(\mathrm{tr}) = \{A \in M_{n\times n}(\mathbb{R}) |\ \mathrm{tr}\, A = 0\} =: \mathfrak{sl}_n(\mathbb{R}),$$

called the *special linear Lie algebra*. Since tr is surjective, we have

$$M_{n\times n}(\mathbb{R})/\mathfrak{sl}_n(\mathbb{R}) \cong \mathbb{R}$$

by the fundamental homomorphism theorem. In particular,

$$\dim\ \mathfrak{sl}_n(\mathbb{R}) = \dim\ M_{n\times n}(\mathbb{R}) - \dim \mathbb{R} = n^2 - 1$$

EXERCISE 2.14. In the alternating group A_4, consider subgroups

$$N = \{1, (1,2)(3,4), (1,3)(2,4), (1,4)(2,3)\} \text{ and } H = \{1, (1,2,3), (1,3,2)\}.$$

(1) Find all left cosets of N in A_4 and check whether the operation on A_4/N defined by $(aN)(bN) = (ab)N$ is well-defined or not.

(2) Check also for A_4/H.

(3) Find a nontrivial homomorphism $\phi : A_4 \to \mathbb{Z}_3$ and show that $A_4/N \cong \mathbb{Z}_3$.

Proposition 2.15. (a) *Let $G = H \times K$ be the direct product of groups H and K. We denote by $\overline{H} = H \times \{e\}$ and $\overline{K} = \{e\} \times K$. Then $\overline{H} \trianglelefteq G$ and $\overline{K} \trianglelefteq G$, with the quotient groups*

$$G/\overline{H} \cong K \quad \text{and } G/\overline{K} \cong H.$$

(b) *Any quotient group of a cyclic group is cyclic.*

(c) *Any quotient of an abelian group is abelian.*

Proof. (a) If we consider the first projection map $\pi_1 : H \times K \to H$, then $\ker \pi_1 = \overline{K}$. So $\overline{K} \trianglelefteq H \times K$ and $(H \times K)/\overline{K} \cong H$ by the fundamental homomorphism theorem. The other follows from $\pi_2 : H \times K \to K$.

(b) Let $N \leq G = \langle a \rangle$. Then $G/N = \langle aN \rangle$.

(c) Let G be abelian and $N \leq G$. Then $(aN)(bN) = (ab)N = (ba)N = (bN)(aN)$ for any $a, b \in G$. □

EXERCISE 2.16. Let N be a normal subgroup of G and H be any subgroup of G.

(1) Note that the smallest subgroup of G containing subgroups N and H is denoted by $N \vee H$. Show that $N \vee H = NH$.
(2) If H is a normal subgroup of G, then prove that NH is normal in G.
(3) Show that $HN/N \cong H/(H \cap N)$.

3. Examples of normal subgroups

EXAMPLE 3.1. (1) $\mathbb{Z}_6/\langle 3 \rangle = \langle 1 + \{0, 3\} \rangle \cong \mathbb{Z}_3$.

(2) $(\mathbb{Z}_4 \times \mathbb{Z}_2)/(\{0\} \times \mathbb{Z}_2) \cong \mathbb{Z}_4$ since we consider $\pi_1 : \mathbb{Z}_4 \times \mathbb{Z}_2 \to \mathbb{Z}_4$ by $(x, y) \mapsto x$ with $\ker \pi_1 = \{0\} \times \mathbb{Z}_2$.

(3) $(\mathbb{Z}_4 \times \mathbb{Z}_6)/\langle (0, 1) \rangle \cong \mathbb{Z}_4$.

(4) $(\mathbb{Z}_4 \times \mathbb{Z}_6)/\langle (0, 2) \rangle \cong \mathbb{Z}_4 \times \mathbb{Z}_2$.

(5) $(\mathbb{Z}_2 \times \mathbb{Z}_4)/\langle (1, 1) \rangle = (\mathbb{Z}_2 \times \mathbb{Z}_4)/\{(0, 0), (1, 1), (0, 2), (1, 3)\} \cong \mathbb{Z}_2$.

(6) $(\mathbb{Z}_4 \times \mathbb{Z}_6)/\langle (2, 3) \rangle = (\mathbb{Z}_4 \times \mathbb{Z}_6)/\{(0, 0), (2, 3)\} \cong \mathbb{Z}_4 \times \mathbb{Z}_3$ because the quotient group $(\mathbb{Z}_4 \times \mathbb{Z}_6)/\langle (2, 3) \rangle$ has an element $(1, 0) + \langle (2, 3) \rangle$ of order 4.

(7) $(\mathbb{Z} \times \mathbb{Z})/\langle (1, 1) \rangle = \langle (1, 0) + \langle (1, 1) \rangle \rangle \cong \mathbb{Z}$.

EXERCISE 3.2. Give the order of the element in the quotient group:

1) $26 + \langle 12 \rangle \in \mathbb{Z}_{60}/\langle 12 \rangle$, 2) $(2, 1) + \langle (1, 1) \rangle \in (\mathbb{Z}_3 \times \mathbb{Z}_6)/\langle (1, 1) \rangle$.

EXAMPLE 3.3. (1) Consider a quotient group $(\mathbb{Z}_4 \times \mathbb{Z}_4 \times \mathbb{Z}_8)/\langle (1, 2, 4) \rangle$. Since the cyclic subgroup is $\langle (1, 2, 4) \rangle = \{(0, 0, 0), (1, 2, 4), (2, 0, 0), (3, 2, 4)\}$, the coset $(0, 0, 1) + \langle (1, 2, 4) \rangle$ is of order 8 in $(\mathbb{Z}_4 \times \mathbb{Z}_4 \times \mathbb{Z}_8)/\langle (1, 2, 4) \rangle$. On the other hand, $(0, 1, 0) + \langle (1, 2, 4) \rangle$ is of order 4. Hence

$$(\mathbb{Z}_4 \times \mathbb{Z}_4 \times \mathbb{Z}_8)/\langle (1, 2, 4) \rangle \cong \mathbb{Z}_4 \times \mathbb{Z}_8.$$

We may also a homomorphism

$$\phi : \mathbb{Z}_4 \times \mathbb{Z}_4 \times \mathbb{Z}_8 \to \mathbb{Z}_4 \times \mathbb{Z}_8 \text{ by } (1, 2, 4) \mapsto (0, 0),\ (0, 1, 0) \mapsto (1, 0),\ (0, 0, 1) \mapsto (0, 1).$$

Then we obtain the above isomorphism by the fundamental homomorphism theorem.

(2) $(\mathbb{Z}\times\mathbb{Z}\times\mathbb{Z})/\langle(3,3,3)\rangle \cong \mathbb{Z}_3\times\mathbb{Z}\times\mathbb{Z}$ because we have an isomorphism mapping $(1,1,1)+\langle(3,3,3)\rangle$ to an element $(1,0,0)$ of order 3, and $(0,1,0)+\langle(3,3,3)\rangle \mapsto (0,1,0)$, $(0,0,1)+\langle(3,3,3)\rangle \mapsto (0,0,1)$. That is, we have a homomorphism

$$\phi:\mathbb{Z}\times\mathbb{Z}\times\mathbb{Z}\to\mathbb{Z}_3\times\mathbb{Z}\times\mathbb{Z} \text{ by } (1,1,1)\mapsto(1,0,0),\ (0,1,0)\mapsto(0,1,0),\ (0,0,1)\mapsto(0,0,1)$$

with $\ker\phi = \mathbb{Z}(3,3,3)$.

EXAMPLE 3.4 (Falsity of the converse of Lagrange's theorem). We show that there is no subgroup H of A_4 such that $|H| = 6$. Suppose that there exists $H \le A_4$ with $|H| = 6 = |A_4|/2$. Then $H \trianglelefteq A_4$ and $A_4/H = \{H, \sigma H\}$ where $\sigma \notin H$. From $HH = H$ and $\sigma H\sigma H = H$, that is, $\alpha^2 \in H$ for $\alpha \in H$ and $\beta^2 \in H$ for $\beta \in \sigma H$, we get that $x^2 \in H$ for $x \in A_4$. So we have more than 6 elements in H:

$$(1,2,3)^2=(1,3,2),\ (1,3,2)^2=(1,2,3),\ (1,2,4)^2=(1,4,2),\ (1,4,2)^2=(1,2,4),$$
$$(1,3,4)^2=(1,4,3),\ (1,4,3)^2=(1,3,4),\ (2,3,4)^2=(2,4,3),\ (2,4,3)^2=(2,3,4),$$

which is a contradiction. Hence there is no subgroup H of A_4 such that $|H| = 6$. In fact,

$$A_4 = \{1,\ (1,2,3),(1,3,2),(1,2,4),(1,4,2),(1,3,4),(1,4,3),(2,3,4),(2,4,3),$$
$$(1,2)(3,4),(1,3)(2,4),(1,4)(2,3)\ \},$$

so A_4 has 4 subgroups of order 3, and a subgroup of order 4, isomorphic to the Klein 4-group.

EXERCISE 3.5. (1) Show that $N = \{1,(1,2)(3,4),(1,3)(2,4),(1,4)(2,3)\}$ is a normal subgroup of A_4, and that $A_4/N = \{N,(1,2,3)N,(1,3,2)N\} \cong \mathbb{Z}_3$.

(2) Show that $N = \{1,(1,2)(3,4),(1,3)(2,4),(1,4)(2,3)\}$ is a normal subgroup of S_4. What is S_4/N? Is it abelian?

EXERCISE 3.6. Let N be a normal subgroup of G and H be any subgroup of G. Let $HN = \{hn|\ h \in H, n \in N\}$. Show that HN is a subgroup of G, which is the smallest subgroup containing both H and N. Furthermore, if H is also normal in G, show that HN is normal in G.

EXERCISE 3.7. Let G be a group and H, K be normal subgroups of G. Show that if $G = \langle H \cup K\rangle$ and $H \cap K = \{e\}$, then $G \cong H \times K$.

Sketch of Solution) To show $G = HK \cong H \times K$, one should check necessarily that for each $g \in G$, there exist unique $h \in H$ and $k \in K$ such that $g = hk$. This means that $\phi : H \times K \to HK$ by $(h, k) \mapsto hk$ is 1-1, or $\psi : HK \to H \times K$ by $hk \mapsto (h, k)$ is well-defined. □

Now we introduce two important examples of normal subgroups.

DEFINITION 3.8. The *center* of a subgroup G is defined to be

$$Z(G) = \{z \in G|\ zg = gz \text{ for all } g \in G\}.$$

Proposition 3.9. (a) *$Z(G)$ is an abelian subgroup of G. Note that $Z(G) = G$ if G is abelian, and that $Z(G) \lneq G$ if G is not abelian.*

(b) *$Z(G)$ is normal in G.*

Proof. The proof of (a) is easy.

(b) For each $g \in G$ and $z \in Z(G)$, $gzg^{-1} = zgg^{-1} = z$, thus $Z(G) \trianglelefteq G$. □

EXAMPLE 3.10. (1) $Z(S_3) = \{\text{id}\}$.

(2) $Z(S_3 \times \mathbb{Z}_5) = \{\text{id}\} \times \mathbb{Z}_5$.

(3) $Z(D_{2\cdot 4}) = \{\text{id}, (1,3)(2,4)\}$.

As a generalization of Example (1), if G is nonabelian of order pq (p, q, distinct primes), then $Z(G)$ is trivial, Why? First, $|Z(G)| \neq pq$ since G is nonabelian. Suppose that $Z(G)$ is of prime order. Then $|G/Z(G)| = |G|/|Z(G)|$ is a prime, so $G/Z(G)$ is cyclic, which is a contradiction to the following proposition.

Proposition 3.11. *If a group G is nonabelian then $G/Z(G)$ is not cyclic. Equivalently, if $G/Z(G)$ is cyclic then G is abelian (and hence $Z(G) = G$).*

Proof. We let $G/Z(G) = \langle aZ(G)\rangle$ for some $a \in G$. Take any two elements $g_1 = a^n z_1$ and $a^m z_2$ in G. Then $g_1 g_2 = a^n z_1 a^m z_2 = a^{n+m} z_2 z_1 = a^m z_2 a^n z_1 = g_2 g_1$. □

EXERCISE 3.12. What is a normal subgroup of A_4? Find $Z(A_4)$.

REMARK 3.13. Later, we will see that a group of order p^n (p, a prime) has a nontrivial center. What is $Z(Q)$, where $Q = \{\pm 1, \pm i, \pm j, \pm k\}$ is the quaternion group?

EXERCISE 3.14. Denote the set of all inner automorphisms of G by $\text{Int}(G)$, and the set of all automorphisms of G by $\text{Aut}(G)$. Note that a subgroup N is normal in G

if N is invariant under every $i_g \in \mathrm{Int}(G)$. The center $Z(G)$ of G consists of fixed elements by every $i_g \in \mathrm{Int}(G)$.

(1) Show that $\mathrm{Int}(G)$ is a normal subgroup of $\mathrm{Aut}(G)$ and $\mathrm{Int}(G)$ is isomorphic to $G/Z(G)$, where $Z(G)$ is the center of a group G.

(2) Show that if G is cyclic then $\mathrm{Aut}(G)$ is abelian, and show that if $\mathrm{Aut}(G)$ is cyclic then G is abelian.

We present another example of normal subgroups.

DEFINITION 3.15. Let G be a group and $a, b \in G$. An element

$$[a, b] = aba^{-1}b^{-1}$$

is called a commutator of the group. The *commutator subgroup* of G is defined by

$$[G, G] = \langle [a, b] = aba^{-1}b^{-1} \mid a, b \in G\rangle.$$

Theorem 3.16. *Let G be a group with commutator subgroup $[G, G]$.*

(a) *The subgroup $[G, G]$ is normal in G*

(b) *The quotient group $G/[G, G]$ is abelian.*

(c) *If N is any normal subgroup of G, then the quotient group G/N is abelian if and only if $[G, G]$ is contained in N. That is, $[G, G]$ is the smallest normal subgroup of G by which the quotient is abelian.*

Proof. (a) We show that for $x \in [G, G]$, $gxg^{-1} \in [G, G]$ for all $g \in G$. We have only to check for any generator $aba^{-1}b^{-1} \in [G, G]$. Since

$$\begin{aligned} g[a, b]g^{-1} &= gaba^{-1}b^{-1}g^{-1} = gaba^{-1}(g^{-1}b^{-1}bg)b^{-1}g^{-1} \\ &= (ga)b(ga)^{-1}b^{-1}(bgb^{-1}g^{-1}) = [ga, b][b, g] \in [G, G], \end{aligned}$$

we proved that $[G, G] \trianglelefteq G$.

(b) For any $a, b \in G$,

$$\begin{aligned} (a[G, G])(b[G, G]) &= (ab)[G, G] = (ab)[b^{-1}, a^{-1}][G, G] \\ &= abb^{-1}a^{-1}ba[G, G] = (ba)[G, G] = (b[G, G])(a[G, G]). \end{aligned}$$

(c) ($\Rightarrow$) Let $a, b \in G$ and G/N be abelian. From $(aN)(bN) = (bN)(aN)$, we have $(aN)(bN)\ (a^{-1}N)(b^{-1}N) = N$, i.e., $(aba^{-1}b^{-1})N = N$. Thus, any commutator $aba^{-1}b^{-1}$ is contained in N.

($\Leftarrow$) Let N be a normal subgroup containing $[G, G]$. Then, for $a, b \in G$,

$$(aN)(bN) = (ab)N = (ab)[b^{-1}, a^{-1}]N = (ba)N = (bN)(aN).$$

Thus we proved the commutativity. □

EXERCISE 3.17. Find $[D_{2\cdot 4}, D_{2\cdot 4}]$ of the group $D_{2\cdot 4}$ of symmetries of a square.

EXAMPLE 3.18. (1) Let us find the commutator subgroup of S_3. Since

$$(1,2,3)(1,2)(1,3,2)(1,2) = (1,3,2) \text{ and } (1,3,2)(1,2)(1,2,3)(1,2) = (1,2,3),$$

we have $A_3 \leq [S_3, S_3]$. But S_3/A_3 is abelian of order 2, we deduce that $[S_3, S_3] = A_3$.

(2) What is the commutator subgroup of S_4? Note that $A_4 = \langle (1,2,3), (1,2)(3,4) \rangle$. From

$$(1,2,3) = (1,3,2)(1,2)(1,2,3)(1,2) \text{ and } (1,2)(3,4) = (1,4,3)(2,4,3)(1,3,4)(2,3,4),$$

we obtain $A_4 \leq [S_4, S_4]$. Thus $[S_4, S_4] = A_4$.

(3) In general, we will see that $[S_n, S_n] = A_n$ for each $n \geq 3$.

EXERCISE 3.19. (1) Check that

$$[A_4, A_4] = \{1, (1,2)(3,4), (1,3)(2,4), (1,4)(2,3)\} \cong \mathbb{Z}_2 \times \mathbb{Z}_2.$$

(2) Show that if N is normal in G then so is $[N, N]$. [6]

REMARK 3.20. * Let X be a topological space and $\pi_1(X)$ be its fundamental group. To obtain the $1st$ *homology group*, we abelianize $\pi_1(X)$, i.e.,

$$H_1(X) = \pi_1(X)/[\pi_1(X), \pi_1(X)].$$

EXERCISE 3.21. * Let G be a group. Given a subgroup H of G, define

$$H_G = \bigcap_{g \in G} g^{-1} H g,$$

called the *core* of H in G.

(1) Prove that if K is a normal subgroup of G contained in H, then K is a subgroup of H_G.

(2) Let $G = GL_2(\mathbb{Q})$ and H be the subgroup of G consisting of nonsingular diagonal matrices. Determine H_G.

[6]Therefore, we obtain the derived series of G. See Ch.3 §3.

Final Exam

(December 16, 2010)

1. Determine whether each statement is true or not: True(T) or False(F).
   (If you answer correctly, you get 3 points; If nothing, 1 point; If wrong, 0 point.)
   1) Every permutation is a cycle.
   2) The dihedral group of the regular n-gon has $2n$ elements.
   3) The symmetric group S_n is generated by adjacent transpositions $(1,2)$, $(2,3)$, $\ldots$, $(n-1,n)$.
   4) The odd permutations in S_n form a subgroup of S_n.
   5) The alternating group A_5 has 120 elements.
   6) The number of left cosets of a subgroup of a finite group divides the order of the group.
   7) If, for a finite group G, $|G| = n$ and d is a divisor of n, then there exists a subgroup $H \leq G$ such that $|H| = d$.
   8) For any subgroup $H \leq G$, it makes sense that we define the operation on the set of left cosets of H by $(aH)(bH) = (ab)H$ for $a, b \in G$.
   9) The subgroup $H = \{1, (1,2)\} \leq S_3$ is a normal subgroup of S_3.
   10) The quotient group $\mathbb{Z}/n\mathbb{Z}$ is cyclic of order n.

2. (10pts) Let $D_{2\cdot 4}$ be the dihedral group of a square. Find all minimal subgroups of $D_{2\cdot 4}$. (Here, a nontrivial subgroup H is called minimal if there is no subgroup K such that $\{1\} \lneq K \lneq H$.)

3. (20pts) Let $\sigma = \begin{pmatrix} 1 & 2 & 3 & 4 & 5 & 6 & 7 & 8 \\ 3 & 5 & 4 & 7 & 2 & 8 & 1 & 6 \end{pmatrix}$ be a permutation in S_8.
   (1) Express σ as a product of disjoint cycles.
   (2) Determine whether σ is even or odd.
   (3) What is the index of the subgroup $\langle \sigma \rangle$ in S_8?
   (4) What is σ^{30}?

4. (10pts) Let $\tau = (1,2,4,3)(5,6)$ and $\sigma = (1,3,4)(2,6)$ be permutations in S_6. Compute σ^{-1} and $\sigma\tau\sigma^{-1}$.

5. (15pts) Find all homomorphisms $\phi : \mathbb{Z}_{12} \to S_3$ and $\ker \phi$.

6. (15pts) Write all elements in the quotient group $\mathbb{Z}_{15}/\langle 3 \rangle$.

4. Simple groups

A longstanding fundamental problem since 19C is the classification of finite simple groups: Find all finite nonabelian simple groups.

DEFINITION 4.1. A group is called *simple* it has no proper nontrivial normal subgroup.

EXAMPLE 4.2. A finite abelian group G simple if and only if $G \cong \mathbb{Z}_p$ for a prime p.

REMARK 4.3. * [Thompson[7] and Feit[8] 1963] If a finite group G is nonabelian and simple, then the order of G must be even.

DEFINITION 4.4. A proper normal subgroup M of G is called maximal if there is no proper normal subgroup N of G such that $M \lneq N \lhd G$.

Lemma 4.5. *Let G and G' be groups and $\phi : G \to G'$ be a homomorphism.*

(a) *If $N \trianglelefteq G$, then $\phi(N) \trianglelefteq \phi(G)$.*

(b) *If $N' \trianglelefteq G'$, then $\phi^{-1}(N') \trianglelefteq G$.*

Proof. (a) If $gng^{-1} \in N$ for any $g \in G$ and $n \in N$, then

$$\phi(gng^{-1}) = \phi(g)\phi(n)\phi(g)^{-1} \in \phi(N),$$

which proves that $\phi(N)$ is normal in $\phi(G)$. Note that, in general, $\phi(N)$ is not normal in G'.

(b) Let $n \in \phi^{-1}(N')$, i.e., $\phi(n) \in N'$. Suppose that $gng^{-1} \notin \phi^{-1}(N')$. Then

$$\phi(g)\phi(n)\phi(g)^{-1} \notin N',$$

which contradicts that $N' \trianglelefteq G'$. Thus $\phi^{-1}(N')$ is normal in G. □

Theorem 4.6. *Let G be a group. Then M is a maximal normal subgroup of G if and only if G/M is simple.*

Proof. Let $\gamma : G \to G/M$ be the canonical projection.

⇒) Suppose that there is a proper nontrivial normal subgroup N' of G/M. Then

$$M \lneq \gamma^{-1}(N') \lhd G,$$

[7]John Griggs Thompson(1932–), an American mathematician.

[8]Walter Feit(1930–2004), a Jewish Austrian-American mathematician who worked in finite group theory and representation theory.

which contradicts that M is maximal.

$\Leftarrow$) Suppose that there exists N such that $M \lneq N \lhd G$. Then

$$\{M\} \lneq \gamma(N) \lhd G/M.$$

This is a contradiction to that G/M is simple. □

EXERCISE 4.7. Find $Z(G)$ and $[G,G]$, where G is a simple nonabelian group.

EXERCISE 4.8 (Jordan[9] 1870). (1) A_n $(n \geq 3)$ contains every 3-cycle.

(2) A_n $(n \geq 3)$ is generated by the 3-cycles.

(3) A_n is generated by the $n-2$ 3-cycles of the form $(1,2,i)$, where $3 \leq i \leq n$.

(4) If a normal subgroup N of A_n $(n \geq 3)$ contains a 3-cycle, then $N = A_n$.

(5) If N is a nontrivial normal subgroup of A_n $(n \geq 5)$, then $N = A_n$.

Solution. (1) $(a,b,c) = (a,b)(a,c) \in A_n$.

(2) Since $(a,b)(c,d) = (a,c,d)(a,c,b)$ and $(a,b)(a,c) = (a,b,c)$, every even permutation is a product of 3-cycles.

(3) Let $3 \leq i,j,k \leq n$ be distinct. Then $(1,2,i)^2 = (1,i,2)$, $(1,2,i)^2(1,2,j) = (1,i,j)$, $(1,2,i)(1,2,j)^2 = (2,i,j)$, and

$$(1,2,i)(1,2,j)^2(1,2,k)(1,2,i)^2 = (i,j,k).$$

Thus, the $n-2$ particular 3-cycles generate all 3-cycles, which proves our assertion by (2).

(4) Assume that $(1,2,3) \in N$ without loss of generality. Then

$$(1,2)(3,j)(1,2,3)^2(3,j)(1,2) = (1,2,j) \in N$$

for $4 \leq j \leq n$. By (3), $N = A_n$.

(5) Take any element of N and divide into several cases as follows:

(Case1: N contains a 3-cycle.)

Assume that $(1,2,3) \in N$ without loss of generality. Then, by (4), $N = A_n$.

(Case2: N contains a product of disjoint cycles, at least one of which has length > 3.)

Let $\sigma = (a_1, a_2, a_3, \ldots, a_r)\mu \in N$. Then $(a_1,a_2,a_3)\sigma(a_1,a_2,a_3)^{-1} \in N$ and

$$(a_1,a_2,a_3)\sigma(a_1,a_2,a_3)^{-1}\sigma^{-1} = (a_2,a_3,a_r) \in N.$$

---

[9]Marie Ennemond Camille Jordan(1838–1922), a French mathematician. He did foundational works in group theory and topology.

Thus, by Case1, $N = A_n$.

(Case3: N contains a disjoint product of the form $\sigma = (a_1, a_2, a_3)(a_4, a_5, a_6)\mu$.)

$$(a_1, a_2, a_4)\sigma(a_1, a_2, a_4)^{-1}\sigma^{-1} = (a_1, a_2, a_4, a_3, a_6) \in N.$$

So, by Case2, $N = A_n$.

(Case4: N contains a disjoint product of the form $\sigma = (a_1, a_2, a_3)\mu$ where μ is a disjoint product of transpositions.)

$\sigma^2 = (a_1, a_3, a_2) \in N$. Thus, by Case1, $N = A_n$.

(Case5: N contains a disjoint product of the form $\sigma = (a_1, a_2)(a_3, a_4)\mu$.)

$(a_1, a_2, a_3)\sigma(a_1, a_2, a_3)^{-1}\sigma^{-1} = (a_1, a_4)(a_2, a_3) \in N$. Let $\alpha = (a_1, a_4)(a_2, a_3)$ and $\beta = (a_1, a_4, i)$. Then $\alpha\beta\alpha^{-1}\beta^{-1} = (a_1, a_4, i) \in N$. So, by Case1, $N = A_n$.

Hence, we have checked for each case of an arbitrary chosen element of N. □

This exercise proves the simplicity of the group A_n $(n \geq 5)$. We have found a series of nonabelian simple groups.

Theorem 4.9. *The group A_n $(n \geq 5)$ is a simple group.*

Now we define another types of groups.

DEFINITION 4.10. A group G is called *solvable* if it has a finite sequence of subgroups $N_0, N_1, \ldots, N_k$ such that

$$\{e\} = N_0 \lhd N_1 \lhd \cdots \lhd N_{k-1} \lhd N_k = G$$

and every quotient N_i/N_{i-1} $(0 \leq i \leq k)$ is abelian.

EXAMPLE 4.11. (1) Every abelian group is solvable.

(2) The dihedral groups D_{2n} are solvable.

(3) The alternating group A_4 is solvable.

(4) Any nonabelian simple group must not be solvable. So, the groups A_n $(n \geq 5)$ are not solvable.

Corollary 4.12. (a) *The only nontrivial proper normal subgroup of S_n $(n \geq 5)$ is A_n.*

(b) *The symmetric group S_n is not solvable for $n \geq 5$.* [10]

[10]The insolvability of S_5 is used to prove that a quintic equation is not solvable by radicals. For details, see Ch.7 §7.

Proof. (b) Note that

$$\{1\} \lhd A_n \lhd S_n,$$

and the alternating group A_n ($n \geq 5$) is a simple nonabelian group. □

REMARK 4.13. * (1) A_5 is the smallest nonabelian simple group, and is the only simple group of order 60 up to isomorphism.[11]

(2) For $n \geq 3$, $[S_n, S_n] = A_n$. The inclusion $\subseteq$ is clear, and every 3-cycle is a commutator of S_n, for instance, $(1,2)(1,2,3)(1,2)^{-1}(1,2,3)^{-1} = (1,2,3)$.

(3) A_n is the only subgroup of S_n having index 2, because a subgroup H of index 2 is normal and a group S_n/H of order 2 is abelian, so $[S_n, S_n] \leq H$.

(4) If $\gamma = (i_0, i_1, \ldots, i_k)$ is a k-cycle in S_n and $\alpha \in S_n$, then $\alpha\gamma\alpha^{-1}$ is also a k-cycle; indeed, $\alpha\gamma\alpha^{-1} = (\alpha(i_0), \alpha(i_1), \ldots, \alpha(i_k))$. Conversely, if $\gamma' = (i'_0, i'_1, \ldots, i'_k)$ is another k-cycle, then there exists $\alpha \in S_n$ with $\gamma' = \alpha\gamma\alpha^{-1}$. In conclusion, all k-cycles are conjugate and two permutations are conjugate if and only if they have the same cycle type.

Since the composition factors are simple groups (we shall study soon), the finite simple groups are the building blocks from which finite groups are constructed.

Theorem 4.14. *The list of finite simple groups is as follows*:

1) $\mathbb{Z}_p$: *cyclic groups of prime order,*
2) A_n ($n \geq 5$): *alternating groups,*
3) $PSL_n(\mathbb{F}_q)$, $P\Omega_n(\mathbb{F}_q)$, $PSp_{2n}(\mathbb{F}_q)$, $\mathsf{G}_2(\mathbb{F}_q)$, … : *simple*[12] *groups of Lie type,*[13]
4) *26 sporadic groups.*[14]

---

[11]See [Hm, Ch.18] and [Hg, Ch.II §8 Exercise 13].

[12]Except for $PSL_2(2)$, $PSL_2(3)$, $P\Omega_5(2)$ and $\mathsf{G}_2(2)$.

[13]These were discovered by Jordan 1870, Dixon 1900, Chevalley 1955, Steinberg 1959, Suzuki 1960. Finally, two families of twisted automorphism groups ${}^2F_4(2^{2n+1})$ and ${}^2G_2(3^{2n+1})$ are known as the *Ree groups*. Rimhak Ree(1922–2005) is the first Korean worldly mathematician, who made distinguished achievements on the classification of finite simple groups in the 1960's. He was one of the first professors of the department of mathematics at Seoul National University, 1946, and published the first article at Bulletin of AMS, 1949, and was a professor at the University of British Columbia, Canada since 1955. He was a member of the Royal Society of Canada. He suffered from Alzheimer's disease in later life and died in January 2005. For details, refer to some articles in *Newsletter of the Korean Math. Soc.* **100** (2005.3), 7–15.

[14]The 5 Mathieu groups, the 4 Janko groups, the 3 Conway groups, the 3 Fischer groups, HS, Mc, Suz, Ru, He, Ly, ON, HN, Th, BM and M. The largest one is called the *monster group*, whose

REMARK 4.15. * (1) Examples of small order among the groups of Lie types are $SL_2(8)$ of order 504 and $PSL_2(7)$ [15] of order 168, etc.

(2) The classification of finite simple groups is believed to be completed by many mathematicians around 1980. "In February 1981 the classification of finite simple groups was completed." So wrote Daniel Gorenstein, the overseer of the programme behind this classification: undoubtedly one of the most extraordinary theorems that pure mathematics has ever seen. The second generation proof appeared in 2004, which was shortened into only 2000 journal pages. For details, see [S, Wi].

5. Subnormal series

* The aim of this section is to introduce the Jordan-Hölder Theorem, which says that every group is expressed as a tower of simple groups.

DEFINITION 5.1. Let G be a group. A finite sequence $\{H_0, H_1, \ldots, H_n\}$ of subgroups of G is a *subnormal series* of G if $\{e\} = H_0 \lhd H_1 \lhd \cdots \lhd H_n = G$.

EXAMPLE 5.2. Consider the 4th dihedral group $D_{2\cdot 4}$. A series

$$\{\mathrm{id}\} \lhd \{\mathrm{id}, (1,2)(3,4)\} \lhd \{\mathrm{id}, (1,3)(2,4), (1,2)(3,4), (1,4)(2,3)\} \lhd D_{2\cdot 4}$$

is a subnormal series of $D_{2\cdot 4}$.

DEFINITION 5.3. Let $\{H_i\}, \{K_j\}$ be two subnormal series of G. We say that $\{K_j\}$ is a refinement of $\{H_i\}$ if $\{H_i\} \subseteq \{K_j\}$. We denote by $\{H_i\} \cong \{K_j\}$ if $\{H_{i+1}/H_i\} \overset{1-1}{\longleftrightarrow} \{K_{j+1}/K_j\}$.

EXAMPLE 5.4. $\{0\} \lhd 72\mathbb{Z} \lhd 24\mathbb{Z} \lhd 8\mathbb{Z} \lhd 4\mathbb{Z} \lhd \mathbb{Z}$ is a refinement of $\{0\} \lhd 72\mathbb{Z} \lhd 8\mathbb{Z} \lhd \mathbb{Z}$.
$\{0\} \lhd \langle 5\rangle \lhd \mathbb{Z}_{15}$ is isomorphic to $\{0\} \lhd \langle 3\rangle \lhd \mathbb{Z}_{15}$.

From the Zassenhaus[16] lemma (so called the butterfly lemma), we obtain:

Theorem 5.5 (Schreier[17]). *Two subnormal series of a group G have isomorphic refinements.*

---

order is approximately $8 \cdot 10^{53}$. This was discovered by R. Griess in 1980 as the automorphism group of the Griess algebra, a 196884-dimensional commutative nonassociative algebra.

[15] $PSL_2(7)$ is the second-smallest nonabelian simple group, next to $A_5 \cong PSL_2(5) \cong SL_2(4)$. It is the automorphism group of the Klein quartic $x^3y + y^3z + z^3x = 0$, which is a Riemann surface and an algebraic curve of genus 3 over $\mathbb{C}$.

[16] Hans Julius Zassenhaus(1912–1991), a German mathematician.

[17] Otto Schreier(1901–1929), an Austrian mathematician

DEFINITION 5.6. A subnormal series $\{H_i\}$ of a group G is called a *composition series* of G if each quotient group H_{i+1}/H_i is simple.

Theorem 5.7 (Jordan[18]-Hölder[19]). *Any two composition series of a group G are isomorphic, which means that the set of factors in a composition series for a group G is an invariant of the group G. In other words, if two groups have different composition factors, then they are non-isomorphic.*

REMARK 5.8. Since the composition factors are simple groups, the finite simple groups are the building blocks from which finite groups are constructed. The complete list of finite simple groups are known (See Theorem 4.14).

DEFINITION 5.9. A group is called *solvable* if there exists a composition series $\{H_i\}$ such that each H_{i+1}/H_i is abelian.

EXAMPLE 5.10. The groups S_n for $n \geq 5$ are not solvable.

EXERCISE 5.11. (1) An abelian group has a composition series if and only if it is finite.

(2) A solvable group with a composition series is finite.

[18]Marie Ennemond Camille Jordan(1838–1922), a French mathematician. He did foundational works in group theory and topology.

[19]Otto Ludwig Hölder(1859–1937), a German mathematician

Final Exam

(December 12, 2013)

1. (30 points) Determine whether each statement is true or not: True(T) or False(F). (If you answer correctly, you get 3 points; If nothing, 1 point; If wrong, 0 point.)
   1) $M_{2\times 2}(\mathbb{R})$ with the multiplication is a commutative group.
   2) For a homomorphism $\phi : \mathbb{Z}_n \to G$, the order of $\phi(1)$ must be a divisor of n.
   3) For a homomorphism $\phi : G \to G'$, $\ker \phi$ is a normal subgroup of G.
   4) The group A_n $(n \geq 3)$ is generated by the 3-cycles.
   5) The group A_4 has a nontrivial proper normal subgroup.
   6) There is a subgroup H of A_4 with $|H| = 6$.
   7) The group A_5 has a nontrivial proper normal subgroup.
   8) The splitting field of $x^3 - 2 \in \mathbb{Q}[x]$ is of dimension 3 as a $\mathbb{Q}$-vector space.
   9) For the splitting field E of a general polynomial of degree 4 in $\mathbb{Q}[x]$, the group $\mathrm{Aut}_{\mathbb{Q}}(E)$ is isomorphic to S_4.
   10) Any polynomial of degree 5 has a root formula using $+$, $\times$ and $\sqrt[n]{\ }$.
2. (30pts) (1) State the definition that a subgroup H of G is normal.
   (2) State the fundamental homomorphism theorem.
   (3) State the definition that a group G is simple.
   (4) State the definition that a group G is solvable.
   (5) Find $Z(D)$, where D is the dihedral group of a square.
   (6) Find $[A_n, A_n]$, where $n \geq 5$.
   - For each of the following problems, you should justify your answer.
3. (20pts) (1) Find all nontrivial homomorphisms $\phi : \mathbb{Z}_9 \to S_3$ with $\ker \phi$.
   (2) Explain why there is no nontrivial homomorphism $\phi : A_5 \to \mathbb{Z}_{30}$.
4. (20pts) (1) Consider a subgroup

$$H = \{1, (1,2,3), (1,3,2)\} \leq A_4.$$

Determine whether the quotient group A_4/H defined by $(aH)(bH) = (ab)H$ for $a, b \in A_4$ is well-defined or not.

(2) Consider the dihedral group D of a square and its subgroup

$$N = \{1, (1,3)(2,4)\}.$$

Among $\mathbb{Z}_4$ and $\mathbb{Z}_2 \times \mathbb{Z}_2$, determine to which group the quotient group D/N is isomorphic.

Final Exam

(December 10, 2012)

1. (30 points) Determine whether each statement is true or not: True(T) or False(F). (If you answer correctly, you get 3 points; If nothing, 1 point; If wrong, 0 point.)
   1) A subgroup H of G is normal in G if $gHg^{-1} = H$ for each $g \in G$.
   2) The subgroup $\{1, (1,2)\}$ of S_3 is normal in S_3.
   3) There is a homomorphism from S_3 with its kernel $\{1, (1,2)\}$.
   4) If a group G has only one subgroup H of a particular order, then H is normal in G.
   5) There is a subgroup $H \leq A_4$ with $|H| = 6$.
   6) The center of a nonabelian group of order pq (p, q, distinct primes) must be trivial.
   7) We get that $[S_n, S_n] = A_n$ $(n \geq 3)$ and $[A_n, A_n] = A_n$ $(n \geq 5)$.
   8) The alternating group A_n $(n \geq 3)$ is a simple group.
   9) There is a nontrivial homomorphism $\phi : A_5 \to \mathbb{Z}_{30}$.
   10) A general equation of degree≥ 5 is solvable by radicals.

2. (20pts) (1) Write down the definition of a solvable group.

   (2) Write down the definition of a simple group.

   (3) Among the following groups, determine which is solvable, and which is simple or not.

$$\mathbb{Z}_5,\ \mathbb{Z}_6,\ D_{2\cdot 4},\ D_{2\cdot 5},\ S_3,\ S_4,\ S_5,\ A_4,\ A_5,\ A_6.$$

• For each of the following problems, you should justify your answer.

3. (20pts) (1) Find all nontrivial homomorphisms $\phi : \mathbb{Z}_4 \to S_3$ with the kernels.

   (2) Find all nontrivial homomorphisms $\phi : \mathbb{Z}_6 \to D_{2\cdot 4}$ with the kernels.

4. (30pts) (1) Consider the dihedral group D of a square and its subgroup

$$N = \{1, (1,3)(2,4)\}.$$

Among $\mathbb{Z}_4$ and $\mathbb{Z}_2 \times \mathbb{Z}_2$, determine to which group the quotient group D/N is isomorphic.

(2) Consider the symmetric group S_4 and its subgroup

$$N = \{1, (1,2)(3,4), (1,3)(2,4), (1,4)(2,3)\}.$$

Among $\mathbb{Z}_6$ and S_3, determine to which group the quotient group S_4/N is isomorphic.

Final Exam

(December 15, 2011)

1. (30 points) Determine whether each statement is true or not: True(T) or False(F). (If you answer correctly, you get 3 points; If nothing, 1 point; If wrong, 0 point.)
   1) The dihedral group D of a square is isomorphic to the quaternion group $Q = \{\pm 1, \pm i, \pm j, \pm k\}$ with $i^2 = j^2 = k^2 = ijk = -1$.
   2) For a group G of prime order, any non-identity element of G is a generator of G.
   3) For a subgroup H of G, $|H|$ is a divisor of $|G|$.
   4) Every subgroup of G is a normal subgroup.
   5) For any homomorphism $\phi : G \to G'$, $\ker \phi$ is a normal subgroup of G.
   6) On the set G/H of left cosets of H in G, the operation $(aH)(bH) = (ab)H$ is well-defined if and only if H is a normal subgroup of G.
   7) In the alternating group A_4 of order 12, there is a subgroup H of order 6.
   8) Any quotient group of an abelian group is abelian.
   9) The alternating group A_n $(n \geq 5)$ is a simple group.
   10) A quintic equation is solvable by radicals, that is, it has a root formula.

- For each of the following problems, you should justify your answer.

2. (5pts) Find all left cosets of $\langle 3 \rangle$ in the group $(\mathbb{Z}_{15}, +)$, and find $(\mathbb{Z}_{15} : \langle 3 \rangle)$.
3. (20pts) (1) Find all nontrivial homomorphisms $\phi : \mathbb{Z}_9 \to S_3$.
   (2) Explain why there is no nontrivial homomorphism $\phi : A_5 \to \mathbb{Z}_{60}$.
   [Hint: $\ker \phi$.]
4. (15pts) Let D be the 4th dihedral group and $N = \{1, (1,3)(2,4)\} \leq D$.
   (1) Show that N is a normal subgroup of D.
   (2) Show that D/N is isomorphic to $\mathbb{Z}_2 \times \mathbb{Z}_2$.
5. (10pts) Let $Q = \{\pm 1, \pm i, \pm j, \pm k\}$ be the quaternion group. Consider the subgroup $N = \{1, -1\}$. Describe the quotient group Q/N, that is, find explicitly the elements of Q/N and the operation table.
6. (20pts) (1) Find all elements in the 4th alternating group A_4.
   (2) Let S_4 be the 4th symmetric group and
   $$N = \{1, (1,2)(3,4), (1,3)(2,4), (1,4)(2,3)\} \trianglelefteq S_4.$$
   Find the elements of S_4/N. To which group is S_4/N isomorphic?

CHAPTER 3

More group theory

In this chapter, we study a little more advanced topics.

1. Group actions

The notion of group actions is fundamental and useful. In particular, it will be used to prove the Sylow theorems in §2.

DEFINITION 1.1. Let G be a group and X be a set. An *action of* G *on* X is a map

$$* : G \times X \to X, \quad (g, x) \mapsto gx$$

such that

i) $ex = x$ for all $x \in X$;
ii) $(g_1 g_2)x = g_1(g_2 x)$ for all $g_1, g_2 \in G$ and all $x \in X$.

We call X to be a *G-set.*

EXAMPLE 1.2. Let X be any set and $H \leq S_X = \{\text{permutations of } X\}$. Then a map $H \times X \to X$ with $(\sigma, x) \mapsto \sigma x = \sigma(x)$ is a group action.

Theorem 1.3. *Let G be a group and X be a G-set. Then*

(a) *The map $\sigma_g : X \to X$ by $x \mapsto gx$ is a permutation of X.*
(b) *The map $\phi : G \to S_X$ by $g \mapsto \sigma_g$ is a group homomorphism.*

Proof. (a) We show that σ_g is 1-1 and onto. Since $gx_1 = gx_2$ for any $x_1, x_2 \in X$ implies that $x_1 = ex_1 = (g^{-1}g)x_1 = g^{-1}(gx_1) = g^{-1}(gx_2) = x_2$, σ_g is 1-1. Also, for any $y \in X$, there is $x = g^{-1}y$ such that $gx = gg^{-1}y = y$, which means that σ_g is onto. So we proved that σ_g is a permutation of X.

(b) We show that $\phi(g_1 g_2) = \sigma_{g_1 g_2} = \sigma_{g_1} \circ \sigma_{g_2} = \phi(g_1) \circ \phi(g_2)$. This follows from the condition of an G-action on X, that is, $\sigma_{g_1 g_2}(x) = (g_1 g_2)x = g_1(g_2 x) = (\sigma_{g_1} \circ \sigma_{g_2})(x)$ for $x \in X$. □

DEFINITION 1.4. Let X be a G-set, that is, $\phi : G \to S_X$ by $g \mapsto \sigma_g$ be a homomorphism. Then the set $N = \phi^{-1}(\text{id})$ of elements leaving every element of X fixed is a normal subgroup of G. So X can be viewed as a G/N-set.

(a) If $N = \{e\}$, then we say that G *acts faithfully* on X.

(b) A group G is called *transitive* on X if for each $x_1, x_2 \in X$, there exists $g \in G$ such that $gx_1 = x_2$. Note that G is transitive on X if and only if the subgroup $\phi(G)$ of S_X is transitive on X.

EXERCISE 1.5. Let X be an nonempty set. Show that if X is finite, then there exists a finite cyclic subgroup H of S_X with $|H| = |X|$ that is transitive on X.

EXAMPLE 1.6. (1) A group G is a G-set since we define an action $G \times G \to G$ by $(g_1, g_2) \mapsto g_1g_2$. For any subgroup $H \leq G$, G is an H-set.

(2) Let $H \leq G$ be a subgroup and L_H be the set of all left cosets of H in G. Then the map $G \times L_H \to L_H$ by $(g, xH) \mapsto (gx)H$ is an well-defined action. Note that every G-set is isomorphic to one that may be formed using these left coset G-sets as building blocks (See [F, §16 Exercises 14–17]).

(3) Let $H \trianglelefteq G$. Then H is a G-set under conjugation, that is, the map $G \times H \to H$ by $(g, h) \mapsto g.h = ghg^{-1}$ is a G-action on H.

(4) Let V be a real vector space and $\mathbb{R}^\times = \mathbb{R} \setminus \{0\}$ be the multiplicative group of nonzero scalars. Then the nonzero scalar multiplication on V, i.e., $\mathbb{R}^\times \times V \to V$ by $(r, v) \mapsto rv$ is an action.

EXERCISE 1.7. Let H be a subgroup of G and let L_H be the set of all left cosets of H.

(1) Show that the map $G \times L_H \to L_H$ given by $(g, xH) \mapsto (gx)H$ is an action.

(2) For any element $g \in G$, show that $\sigma_g : L_H \to L_H$ given by $\sigma_g(xH) = (gx)H$ is a permutation on L_H.

(3) Let $S(L_H)$ be the group of permutations on L_H. Show that $\sigma : G \to S(L_H)$ given by $\sigma(g) = \sigma_g$ is a group homomorphism.

REMARK 1.8. * Instead of S_X in Theorem 1.3, if we take $GL(V)$ where V is an n-dimensional vector space, then we obtain an important object.

More precisely, let $GL_n(\mathbb{C}) = \{A \in M_{n\times n}(\mathbb{C}) |\ A \text{ is invertible}\}$ be the general linear group of degree n over $\mathbb{C}$. A group homomorphism

$$\phi : G \longrightarrow GL_n(\mathbb{C})$$

is called a (linear or matrix) *representation* of G of degree n over $\mathbb{C}$. Note that this is a G-action on an n-dimensional vector space $V \cong \mathbb{C}^n$.[1] We call V to be a *G-module.*

Proposition 1.9. *Let X be a G-set and $x \in X$.*

(a) *We define a subset of elements in G under whose actions $x \in X$ remains fixed, i.e.,*

$$G_x = \{g \in G | \ gx = x\}.$$

Then G_x is a subgroup of G, called the isotropy subgroup *of $x \in X$. Also, we denote by $X_g = \{x \in X | \ gx = x\}$.*

(b) *A relation $\sim$ on X is defined by*

$$x_1 \sim x_2 \ \Leftrightarrow \ gx_1 = x_2 \ \textit{ for some } g \in G.$$

Then this is an equivalence relation. The equivalence class $[x]$ containing x is called the orbit of $x \in X$ under G, *denoted by Gx.*

Proof. (a) For each $g_1, g_2 \in G_x$, $(g_1g_2)x = g_1(g_2x) = g_1x = x$, so $g_1g_2 \in G_x$. Clearly, $e \in G_x$ since $ex = x$. For each $g \in G_x$, $x = ex = (g^{-1}g)x = g^{-1}(gx) = g^{-1}x$, so $g^{-1} \in G_x$.

(b) The relation $\sim$ is reflexive($x \sim x$), symmetric($x \sim y \Rightarrow y \sim x$) and transitive ($x \sim y$ and $y \sim z \Rightarrow x \sim z$). □

Theorem 1.10. *Let a group G act on a set X and $x \in X$. Then the cardinality $|Gx|$ of the G-orbit of x is the same as the index $(G : G_x)$ of the stabilizer G_x in G. In particular, for a finite group G, $|Gx|$ is a divisor of $|G|$, i.e.,*

$$|Gx| = (G : G_x) = |G|/|G_x|.$$

Proof. Let G/G_x be the set of all left cosets of G_x in G. Define $\phi : G/G_x \to Gx$ by $\phi(gG_x) = gx$. Then we show that ϕ is a well-defined bijection. If $g_1G_x = g_2G_x$, i.e., $g_1^{-1}g_2 \in G_x$, then $g_1x = g_1(g_1^{-1}g_2x) = g_2x$, so the map ϕ is well-defined. The proof that ϕ is 1-1 and onto is left as an exercise. □

EXAMPLE 1.11. Consider the 4-th dihedral group $D_{2\cdot 4}$, which is the group of symmetries of a square. Let X be the set consisting of vertices $1, 2, 3, 4$, edges e_1, e_2, e_3, e_4,

[1]It is possible to take another field, for example, $\mathbb{Q}, \mathbb{R}$ or the algebraic closure $\overline{\mathbb{F}_p}$ of the finite field $\mathbb{F}_p$ (p, a prime). Of course, the situations are much more complicated.

vertical and horizontal axes m_1, m_2, the diagonals d_1, d_2, the center point C, the midpoints P_1, P_2, P_3, P_4 of the edges in a square. Then X is a $D_{2\cdot 4}$-set.

Let us find an orbit, an isotropy subgroup, and a set of fixed elements in X. E.g., $D_{2\cdot 4}1 = \{1, 2, 3, 4\}$, $(D_{2\cdot 4})_1 = \{\mathrm{id}, (2, 4)\}$ and $X_{(1,2)(3,4)} = \{e_1, e_3, m_1, m_2, C, P_1, P_3\}$.

Proposition 1.12. *For a G-set X, let $g \in G$ and $x \in X$. Then*

$$\{h \in G |\ hx = gx\} = gG_x.$$

Proof. Left as an exercise. □

Theorem 1.13 (Burnside's[2] formula). *Let G be a finite group and X be a finite G-set. If we let r to be the number of orbits in X under G, then*

$$r\ |G| = \sum_{g \in G} |X_g|.$$

Proof. Recall that $X_g = \{x \in X |\ gx = x\}$. We let $S = \{(g, x) |\ gx = x\}$. Then $|S| = \sum_{g \in G} |X_g|$, or

$$|S| = \sum_{x \in X} |G_x| = \sum_{x \in X} \frac{|G|}{|Gx|} = |G| \sum_{x \in X} \frac{1}{|Gx|} = |G| \sum_{O,\text{orbits}} \sum_{x \in O} \frac{1}{|Gx|} = |G| \sum_{O,\text{orbits}} 1 = |G| r.$$

□

Corollary 1.14. *With the same notations as above,*

$$(\text{the number of orbits}) = \frac{\sum_{g \in G} |X_g|}{|G|} = \frac{\sum_{x \in X} |G_x|}{|G|}.$$

Exercise 1.15. (1) Show that the number of cases that 7 persons sit around a round table is $7!/7$.

(2) Show that the number of necklaces which are threaded with 7 distinct beads is $7!/14$.

(3) Show that the number of colorings that each side of a regular triangle is painted with one of 4 colors is $(4^3 + 4 + 4 + 4^2 + 4^2 + 4^2)/6 = 20$.

---

[2]William Burnside(1852–1927), an English mathematician.

2. Sylow's theorems

Consider a G-set X. Recall that $Gx = \{gx|\ g \in G\}$. We let r be the number of orbits, and $\{x_1, x_2, \ldots, x_r\}$ be the set of representatives from all orbits in X. There may be one-element orbits in X. Let $X_G = \{x \in X|\ gx = x \text{ for each } g \in G\} = \{x_1, x_2, \ldots, x_s\}$ where $0 \leq s \leq r$. Then

$$|X| = \sum_{i=1}^{r} |Gx_i| = |X_G| + \sum_{i=s+1}^{r} |Gx_i|.$$

Lemma 2.1. *Let G be a group with $|G| = p^n$ (p, a prime), and X be a finite G-set. Then $|X| \equiv |X_G| \pmod p$.*

Proof. Note that $|X| = |X_G| + \sum_{|\text{orbit}|>1} |Gx_i|$. Since $1 < |Gx_i| = (G : G_{x_i})$ and $(G : G_{x_i})G_{x_i} = |G| = p^n$, we get that $p|\ |Gx_i|$, thus $p|(|X| - |X_G|)$. □

DEFINITION 2.2. Let p be a prime. A group G is called a *p-group* if each element has an order of a power of p, that is, for each $g \in G$, $|g| := |\langle g\rangle| = p^r$ $(r \in \mathbb{Z}_{\geq 0})$.

Theorem 2.3 (Cauchy[3]). *Let G be a finite group and its order is divisible by a prime p. Then G has an element of order p.*

Proof. Let $X = \{(g_1, g_2, \ldots, g_p)|\ g_i \in G, g_1g_2\cdots g_p = e\}$. Then $|X| = |G|^{p-1}$ because $g_p = (g_1g_2\cdots g_{p-1})^{-1}$, so $p|\ |X|$. Take a p-cycle $\sigma = (1, 2, \ldots, p) \in S_p$, and define

$$\sigma(g_1, g_2, \ldots, g_p) = (g_2, g_3, \ldots, g_p, g_1) \in X.$$

Then we have a group action $\langle\sigma\rangle \times X \to X$. Since $|\langle\sigma\rangle| = p$, $|X| \equiv |X_{\langle\sigma\rangle}|$ by the previous Lemma. So $p|\ |X_{\langle\sigma\rangle}|$. We note that $(g_1, g_2, \ldots, g_p) \in X_{\langle\sigma\rangle}$ if and and only if

$$g_1 = g_2 = \cdots = g_p.$$

Clearly, $(e, e, \ldots, e) \in X_{\langle\sigma\rangle}$, and the fact that $p|\ |X_{\langle\sigma\rangle}|$ leads to the existence of a non-identity $g \in G$ such that $(g, g, \ldots, g) \in X_{\langle\sigma\rangle}$. Hence $g^p = e$ and $\langle g\rangle$ is a subgroup of order p. □

Corollary 2.4. *Let $|G| < \infty$. Then G is a p-group if and only if $|G| = p^n$ for some n.*

---

[3]Augustin-Louis Cauchy(1789–1857), a French mathematician. He was an early pioneer of Analysis.

For the statements and proofs of the Sylow[4] theorems, we consider the following setting. Let G be a group and let $\mathscr{S} = \{H|\ H \leq G\}$. We define a group action $G \times \mathscr{S} \to \mathscr{S}$ by conjugation, i.e., $(g, H) \mapsto gHg^{-1}$. Then the isotropy subgroup of H is

$$G_H = \{g \in G|\ gHg^{-1} = H\} \leq G.$$

Note that $H \trianglelefteq G_H$, and G_H is the largest subgroup of G having H as a normal subgroup, which is called the *normalizer* of H in G, denoted by $N[H]$.

Lemma 2.5. *If H is a p-subgroup of a finite group G, then*

$$(N[H] : H) \equiv (G : H) \pmod{p}.$$

Proof. Let $\mathscr{L} = \{gH|\ g \in G\}$ with $|\mathscr{L}| = (G : H)$. Consider a group action $H \times \mathscr{L} \to \mathscr{L}$ by $(h, xH) \mapsto (hx)H$. Then the set of each left coset which is fixed under all actions of H is

$$\mathscr{L}_H = \{xH|\ h(xH) = xH \text{ for all } h \in H\} = \{xH|\ x^{-1}hx \in H \text{ for all } h \in H\}.$$

Note that $xH \in \mathscr{L}_H$ is the same as $x \in N[H]$. So the number of left cosets in $\mathscr{L}_H$ is $|\mathscr{L}_H| = (N[H] : H)$. By Lemma 2.1, $|\mathscr{L}_H| \equiv |\mathscr{L}| \pmod{p}$. ☐

Corollary 2.6. *Let H be a p-subgroup of a finite group G. If $p|(G : H)$, then $N[H] \neq H$.*

Theorem 2.7 (Sylow 1st). *Let G be a finite group and $p^n \| \,|G|$, i.e., $p^n|\,|G|$ but $p^{n+1} \nmid |G|$ for a prime p. Then*

(a) *There exist subgroups H_i of G such that $|H_i| = p^i$, where $1 \leq i \leq n$.*

(b) *Moreover, $\{e\} \lhd H_1 \lhd H_2 \lhd \cdots \lhd H_n \leq G$.*

Proof. (a) The case of $i = 1$ is Cauchy's theorem. We use induction on i. Assume that the cases for $i < n$ hold. If we let H be a subgroup of order p^i, then $p|(G : H)$. By Lemma 2.5, $p|(N[H] : H)$. Since $H \lhd N[H]$, we have the quotient group $N[H]/H$, and $p|\,|N[H]/H|$. By Cauchy's theorem, there is a subgroup K in $N[H]/H$ with $|K| = p$. If we consider the canonical epimorphism $\gamma : N[H] \to N[H]/H$, then

$$H \lneq \gamma^{-1}(K) = \{x \in N[H]|\ \gamma(x) \in K\} \leq N[H] \leq G.$$

Hence $|\gamma^{-1}(K)| = p^{i+1}$.

[4] Peter Ludwig Mejdell Sylow(1832–1918), a Norwegian mathematician.

(b) Note that $H \lneq \gamma^{-1}(K) \leq N[H]$ in (a). Since $H \lhd N[H]$, the subgroup H is normal in $\gamma^{-1}(K)$. □

DEFINITION 2.8. A maximal p-subgroup of a group G is called a *Sylow p-subgroup* of G.

Theorem 2.9 (Sylow 2nd). *Let P_1 and P_2 be Sylow p-subgroups of a finite group G. Then P_1 and P_2 are conjugate in G.*

Proof. Let $\mathscr{L} = \{gP_1 | \ g \in G\}$ be a P_2-set by $(y, xP_1) \mapsto (yx)P_1$. Note that $|\mathscr{L}| = (G : P_1)$ is not divisible by p. From Lemma 2.1, $|\mathscr{L}| \equiv |\mathscr{L}_{P_2}| \pmod p$, so $|\mathscr{L}_{P_2}| \neq 0$. Thus there exists $x \in G$ such that $yxP_1 = xP_1$ for any $y \in P_2$. That is, $x^{-1}yx \in P_1$ for any $y \in P_2$. Since $x^{-1}P_2x \leq P_1$ and $|P_1| = |P_2|$, we have $P_1 = x^{-1}P_2x$. □

Theorem 2.10 (Sylow 3rd). *Let G be a finite group and $p| \, |G|$. Then the number n_p of Sylow p-subgroups of G is congruent to 1 modulo p and divides $|G|$.*

Proof. Let P be a Sylow p-subgroup and $\mathscr{S}$ be the set of all p-subgroups. We define a group action of P on $\mathscr{S}$ by conjugation, i.e., $(x, S) \mapsto xSx^{-1}$. By Lemma 2.1, we know that $|\mathscr{S}| \equiv |\mathscr{S}_P| \pmod p$. Note that $S \in \mathscr{S}_P$ means that $xSx^{-1} = S$ for all $x \in P$. Thus $P \leq N[S]$ and clearly $S \trianglelefteq N[S]$, so both P and S are Sylow p-subgroups of $N[S] \leq G$. Since P and S are conjugate in $N[S]$ by the 2nd Sylow theorem, the fact that $S \trianglelefteq N[S]$ leads to $P = S$, i.e., $\mathscr{S}_P = \{P\}$ and $|\mathscr{S}| \equiv 1 \pmod p$.

On the other hand, we consider a G-set $\mathscr{S}$ by conjugation. By the 2nd Sylow theorem, $\mathscr{S}$ has only one orbit. Hence, for $P \in \mathscr{S}$, $|\mathscr{S}| = |G.P| = (G : G_P)$, which is a divisor of $|G|$. □

EXAMPLE 2.11. (1) The symmetric group S_3 is of order $6 = 2 \times 3$. The order of a Sylow 2-subgroup of S_3 is 2. The subgroups of order 2 are $\{\mathrm{id}, (1,2)\}$, $\{\mathrm{id}, (2,3)\}$, $\{\mathrm{id}, (1,3)\}$. The number n_2 of Sylow 2-subgroups is 3, which is congruent to 1 modulo 2 and divides $|S_3| = 6$. The 3 Sylow 2-subgroups are all conjugate.

(2) Let G be a group of order $15 = 3 \times 5$. Then we claim that G must not be simple. First, there is a subgroup H of order 5. By the 3rd Sylow theorem, $n_5 \equiv 1 \pmod 5$ and $n_5 | 15$, thus $n_5 = 1$ and $gHg^{-1} = H$ for all $g \in G$. Hence we see that H is a normal subgroup of G, which proves our claim. This example is generalized to the case of order pq, where p and q are distinct primes.

We give some applications of the Sylow theorems.

Theorem 2.12. *Any finite p-group is solvable.*

Proof. Let G be a group of order p^n. By the 1st Sylow theorem, there exist subgroups H_i of order p^i and $H_i \lhd H_{i+1}$. Then we have a composition series

$$\{e\} \lhd H_1 \lhd H_2 \lhd \cdots \lhd H_n = G.$$

Note that $|H_{i+1}/H_i| = p$, so H_{i+1}/H_i is cyclic, thus abelian. □

Theorem 2.13. *The center of a nontrivial finite p-group G contains more than one element.*

Proof. Consider the *class equation* of G coming from the action of G on G by conjugation,

$$|G| = |Z(G)| + \sum(G : Z_G(x)),$$

the sum being taken over certain x for which $(G : Z_G(x)) \neq 1$. Since $|G|$ and each $(G : Z_G(x))$ are divisible by p, so is $|Z(G)|$. Hence $|Z(G)| \geq p$. □

Remark 2.14. * There is an infinite p-group which has a trivial center.

Exercise 2.15. (1) Show that every group of order p^2 is abelian. How about a group of order p^3?

(2) Let G be a nonabelian group of order p^3. Determine the order of the center $Z(G)$ and describe $G/Z(G)$ up to isomorphism.
[Hint: If $G/Z(G)$ is cyclic then G is abelian.]

(3) The two nonabelian groups of order 8 are $D_{2\cdot 4} = \langle a, b|\ a^4 = b^2 = 1,\ ba = a^3b\rangle$ and $Q = \langle x, y|\ x^4 = 1,\ x^2 = y^2,\ yx = x^3y\rangle$. Show that

$$Z(D_{2\cdot 4}) = \langle a^2\rangle \cong \mathbb{Z}_2 \cong \langle x^2\rangle = Z(Q)$$

and deduce that

$$\mathrm{Int}(D_{2\cdot 4}) \cong \mathbb{Z}_2 \times \mathbb{Z}_2 \cong \mathrm{Int}(Q).$$

Exercise 2.16. Let G be a group of order pq, where p, q are primes with $p > q$.

(1) Prove that if $q \nmid p - 1$ then G is abelian.

(2) Prove that if $q = 2$ then G is isomorphic either to the cyclic group $\mathbb{Z}_{2p}$ or the dihedral group D_{2p}.

Exercise 2.17. Show that the number of distinct conjugates of H in G is $[G : N(H)]$.

Exercise 2.18. (1) Prove that all group of order $5^2 \cdot 7^2$ is abelian.

(2) Prove that no group of order 48 is simple.

(3) Prove that any group of order $5 \cdot 7 \cdot 47$ is cyclic.

Exercise 2.19. Show that no group of order 200 is simple. [Hint: Consider the number of Sylow 5-groups.]

Exercise 2.20 (Finite general linear group). *

(1) Prove that there is a Sylow p-subgroup T of $GL_n(\mathbb{F}_p)$ such that every element of T is an upper triangular matrix, where $\mathbb{F}_p$ is the field of p elements where p is a prime. Find the order of T. [Hint: $|GL_n(\mathbb{F}_p)| = (p^n - 1)(p^n - p) \cdots (p^n - p^{n-1})$.]

(2) Let $\mathcal{S}$ be the set of all Sylow p-subgroups of $GL_2(\mathbb{F}_p) =: G$. Let G act on $\mathcal{S}$ by conjugation. Find the stabilizer(isotropy subgroup) and the orbit of H, where $H = \left\{ \begin{pmatrix} 1 & b \\ 0 & 1 \end{pmatrix} \mid b \in \mathbb{F}_p \right\}$.

(3) Find the number of Sylow p-subgroups of G.

[Hint: A group action gives $|G.H| = |G|/|G_H|$.]

Final Exam

(December 14, 2009)

1. Determine whether each statement is true or not: True(T) or False(F).
   (If you answer correctly, you get 3 points; If nothing, 1 point; If wrong, 0 point.)
   1) A homomorphism is 1-1 if and only if the kernel consists of only the identity.
   2) A homomorphism ϕ from a cyclic group $G = \langle g \rangle$ is determined by the value $\phi(g)$.
   3) Let H be a normal subgroup of K and K be a normal subgroup of G. Then H is a normal subgroup of G.
   4) The alternating group A_n $(n \geq 5)$ has a nontrivial normal subgroup.
   5) A Sylow p-subgroup of a finite group G is normal in G if and only if it is the only Sylow p-subgroup of G.
2. (15 pts) Let

$$G = \{\text{id}, (1,2,3,4), (1,3)(2,4), (1,4,3,2), (1,2)(3,4), (1,4)(2,3), (1,3), (2,4)\}$$

   be the 4th dihedral group and X be the G-set consisting of vertices $1, 2, 3, 4$, edges e_1, e_2, e_3, e_4, vertical and horizontal axes m_1, m_2, the diagonals d_1, d_2, the center point C, the midpoints P_1, P_2, P_3, P_4 of the edges in a square.
   (1) Find the orbit Ge_1 under G.
   (2) Find the isotropy subgroup G_{e_1} and the index $(G : G_{e_1})$.
   (3) Find the fixed set $X_{(1,3)(2,4)}$.

• For each of the following problems, you should justify your answer.

3. (20 pts) (1) Find the number of distinct group homomorphisms from $\mathbb{Z}_2 \times \mathbb{Z}_3$ into the symmetric group S_3.
   (2) Find all nontrivial homomorphisms $\phi : \mathbb{Z}_{12} \to \mathbb{Z}_{12}$. What is the kernel of each nontrivial homomorphism?
4. (30 pts) Classify the given groups according to the fundamental theorem of finitely generated abelian groups, that is, find a finite abelian group isomorphic to:
   (1) $(\mathbb{Z}_9 \times \mathbb{Z}_5)/\langle(3,0)\rangle$, (2) $(\mathbb{Z}_{12} \times \mathbb{Z}_{18})/\langle(4,3)\rangle$, (3) $(\mathbb{Z} \times \mathbb{Z})/\langle(2,2)\rangle$.
5. (10 pts) Show that a finite simple abelian group is isomorphic to $\mathbb{Z}_p$ for a prime p.

6. (20 pts) (1) Let G be a simple nonabelian group. Show that the center of G is trivial.

   (2) Find the commutator subgroup of the symmetric group S_3.

7. (20 pts) Let H be a subgroup of G and let L_H be the set of all left cosets of H.

   (1) For any element $g \in G$, show that $\sigma_g : L_H \to L_H$ given by $\sigma_g(xH) = (gx)H$ is a permutation on L_H. [Hint: Show that it is 1-1 and onto.]

   (2) Let $S(L_H)$ be the group of permutations on L_H. Show that $\sigma : G \to S(L_H)$ given by $\sigma(g) = \sigma_g$ is a homomorphism. Equivalently, you may show that the map $G \times L_H \to L_H$ given by $(g, xH) \mapsto (gx)H$ is an action.

8. (15 pts) Show that there are no simple groups of order $2 \cdot 3^n$ (n, a positive integer). [Hint: Calculate the number of Sylow 3-subgroups.]

3. Nilpotent and solvable groups *

We study more thoroughly an important class of groups.

3.1. Nilpotent groups.

Theorem 3.1. *A finite group G is* nilpotent *if the following equivalent conditions hold*:

(a) $\langle e\rangle \leq Z_1(G) \leq Z_2(G) \leq \cdots \leq Z_n(G) = G$ *where and* $Z_1(G) = Z(G)$ *and*

$$Z_{i+1}(G)/Z_i(G) = Z(G/Z_i(G)),$$

that is, the upper central series of G terminates in G. Note that the ith central subgroup $Z_i(G)$ is a characteristic subgroup of G.

(b) $G = G^0 \geq G^1 \geq G^2 \geq \cdots \geq G^n = \langle e\rangle$ *where*

$$G^{i+1} = [G, G^i] = \langle aba^{-1}b^{-1} |\ a \in G, b \in G^i\rangle,$$

that is the lower central series of G terminates in $\langle e\rangle$.

(c) *There is a normal series $G = G^0 \geq G^1 \geq G^2 \geq \cdots \geq G^n = \langle e\rangle$ such that $[G, G^i] \leq G^{i+1}$ for $0 \leq i \leq n-1$, that is, G can be constructed by central extensions.*

(d) *G is the direct product of its Sylow subgroups, i.e., every Sylow subgroup is normal in G.*

(e) (Wielandt[5]) *Every maximal subgroup of G is normal.*

(f) *Every proper subgroup of G is properly contained in its normalizer.*

Example 3.2. A finite p-group is nilpotent.

Proposition 3.3. (a) *Every subgroup, and every quotient of a nilpotent group is nilpotent.*

(b) *A group G is nilpotent if and only if $G/Z(G)$ is nilpotent.*

3.2. Solvable groups.

Theorem 3.4. *A finite group G is* solvable *if the following equivalent conditions hold*:

---

[5]Helmut Wielandt(1910–2001), a German mathematician.

(a) $G = G^{(0)} \geq G^{(1)} \geq G^{(2)} \geq \cdots \geq G^{(n)} = \langle e \rangle$ *where*

$$G^{(i+1)} = [G^{(i)}, G^{(i)}] = \langle aba^{-1}b^{-1} \mid a, b \in G^{(i)} \rangle,$$

that is, the derived series of G *terminates in* $\langle e \rangle$. *Note that the ith derived subgroup* $G^{(i)}$ *is a fully invariant subgroup of* G.

(b) G *has a subnormal series whose factors are all abelian, that is,* G *has a solvable series.*

(c) G *has a composition series whose factors are all cyclic, that is,* G *can be cyclic extensions.*

(d) (Hall[6]) *If* $|G| = mn$ *and* $(m, n) = 1$, *then* G *has a subgroup of order* m.

(e) (Thompson) *If for any pair of elements* $x, y \in G$, *the derived series of* $\langle x, y \rangle$ *terminates in* $\langle e \rangle$.

EXERCISE 3.5. Find a composition series of S_4. Is S_4 solvable? And, is S_5 solvable?

Proposition 3.6. (a) *Every nilpotent group is solvable.*

(b) *Every subgroup, every quotient, and every homomorphic image of a solvable group is solvable.*

(c) *If* N *and* G/N *are solvable, then so is* G.

EXERCISE 3.7. Show that if H and K are solvable normal subgroups of a group G then so is $H + K$.

EXERCISE 3.8. (1) Prove that every subgroup H of a finite group G is normal in G if $(G : H)$ is the smallest prime dividing $|G|$, the order of G.

(2) Prove that every group of order 1995 that contains a subgroup of order 665 is solvable. [Hint: Use (1) to find a composition series.]

EXERCISE 3.9. (1) Show that any group of order p^2q is solvable.

(2) Let $p < q < r$ be primes. Show that a group of order pqr is solvable.

Proposition 3.10. *Any group of order less than 60 is solvable.*

Theorem 3.11 (Burnside's p^aq^b theorem). [7] *Any group of order* p^aq^b *is solvable.*

---

[6]Philip Hall(1904–1982), an English mathematician.

[7]W. Burside, *On groups of order* $p^\alpha q^\beta$, Proc. London Math. Soc. **S2-1** (1904), 388–392.

REMARK 3.12. The proof given by Burnside in 1904 was using the arithmetic of group characters in group representation theory. Later, an elementary proof was set down by Bender in 1972.

Theorem 3.13 (Feit-Thompson). *Any group of odd order is solvable.*

Corollary 3.14. *Every nonabelian simple group must be of even order.*

Theorem 3.15 (Iwasawa[8]). * *Suppose that G acts on a set S faithfully and primitively*[9]. *Fix $s \in S$ and let G_s be the stabilizer of s. Suppose there is a solvable subgroup $H \trianglelefteq G_s$ such that $G = \bigcup\{H^x = xHx^{-1} \mid x \in G\}$. Then G is simple.*

Proof. Suppose that $\langle e \rangle \neq N \trianglelefteq G$ is transitive on S, and $G = G_sN$. Thus $HN \trianglelefteq G_sN = G$. If $x \in G$ then $H^x \leq (HN)^x = HN$, so $\bigcup\{H^x \mid x \in G\} \subseteq HN$ and hence $HN = G$. Since H is solvable, $H^m = \langle e \rangle$ for some H^m in the derived series of H. Check inductively that $(HN)^l \leq H^l$ for all l. Thus $G = G^m = (HN)^m N = N$ and $N = G$. □

4. Free abelian groups

Theorem 4.1. *Let F be a free abelian group of rank n, and G a subgroup of F. Then G is a free abelian group of rank $m \leq n$.*

REMARK 4.2. The theorem holds even if the rank is infinite.

EXERCISE 4.3. Let G be a finitely generated abelian group in which no element except 0 has finite order. (That is, G a finitely generated torsion-free abelian group.) Show that G is a free abelian group.

EXERCISE 4.4. (1) The additive group $\mathbb{Q}$ is not finitely generated, and $\mathbb{Q}$ is not a free abelian group.

(2) Let A be the subgroup of $(\mathbb{R}^2, +)$ generated by $(1,0), (\sqrt{2}, 0)$. Is A a free abelian group?

EXERCISE 4.5. Let G be an additive abelian group with 3 generators and relations

$$a + 2c = 0, \;\; 3a + 3b + 9c = 0.$$

Write down G as the direct product of cyclic groups.

---

[8]Kenkichi Iwasawa(1917–1998), a Japanese mathematician, known for his influence on algebraic number theory.

[9]A group action is called primitive if it is transitive and it has no nontrivial group blocks.

5. Free groups and free products *

We introduce a language from the category theory and apply to the group theory.

Theorem 5.1 (Category of groups).

(a) *Let* $\{G_i\}$ *be a family of groups.* $\prod G_i$ *is a product in the category of groups.*

(b) *Let* $\{G_i\}$ *be a family of abelian groups.* $\bigoplus G_i$ *is a coproduct in the category of abelian groups.*

EXAMPLE 5.2 (Amalgamated product). *

Let G, G_1, G_2 be groups with $f_i : G \to G_i$, homomorphisms for $i = 1, 2$.

F:=the free group generated by $G_1 \amalg G_2$. Denote by $x \cdot y$ the product in F.

R:=the normal subgroup of F generated by the words $f_1(z) \cdot f_2(z)^{-1}$ for $z \in G$.

The amalgamated product of G_1 and G_2 over G, $G_1 *_G G_2 := F/R$.

Suppose $X = U \cup V$ with $x_0 \in U \cup V$ where U, V and $U \cap V$ are open and path-connected topological spaces. Then $\pi_1(X, x_0) \cong \pi_1(U, x_0) *_{\pi_1(U \cap V, x_0)} \pi_1(V, x_0)$ where the homomorphisms for an amalgamated product are induced by inclusions, which is the *Seifert-van Kampen theorem* in algebraic topology.

REMARK 5.3. Note that π_1 is a covariant functor from the category of topological spaces to the category of groups, and the free product $\prod^* G_i$ is the coproduct(i.e., the initial object) in the category of groups.

EXERCISE 5.4. Let F be a free group on a set S.

(1) Show that the commutator group $[F, F]$ is a normal subgroup of F.

(2) Describe the quotient group $F/[F, F]$.

6. Some finite groups

EXERCISE 6.1. Classify all finite groups of order 8.

Solution. We know that three abelian ones:

$$\mathbb{Z}_8,\ \mathbb{Z}_2 \times \mathbb{Z}_4,\ \mathbb{Z}_2 \times \mathbb{Z}_2 \times \mathbb{Z}_2.$$

Let G be a nonabelian group of order 8. Each element but the identity is of order 2 or 4. If every element were of order 2, then G is abelian, contradiction. Let $\langle a \rangle$ be a subgroup of order 4 and then $\langle a \rangle \lhd G$. If $b \notin \langle a \rangle$, $\langle a \rangle \cup b\langle a \rangle$ is the coset decomposition of G. Since $G/\langle a \rangle \cong \mathbb{Z}_2$, $b^2 \in \langle a \rangle$. If $b^2 = a$ or $b^2 = a^3$, then b would be of order

8. So $b^2 = 1$ or $b^2 = a^2$. Since $b\langle a\rangle b^{-1} = \langle a\rangle$, we have $bab^{-1} = a^3$. Thus two possibilities for G are

$$G_1 = \langle a, b |\ a^4 = 1,\ b^2 = 1,\ bab^{-1} = a^3\rangle,\ G_2 = \langle a, b |\ a^4 = 1,\ b^2 = a^2,\ bab^{-1} = a^3\rangle,$$

whose elements are $1, a, a^2, a^3, b, ab, a^2b, a^3b$. □

REMARK 6.2. The above G_1 is the dihedral group of degree 4 and G_2 is called the quaternion group. Note that

$$G_2 = \langle i, j |\ i^4 = 1,\ j^2 = i^2 = -1,\ ji = -ij\rangle = \{\pm 1, \pm i, \pm j, \pm ij\}.$$

EXERCISE 6.3. * Let G be the group given by $G = \langle a, b |\ a^2, b^2, (ab)^3\rangle$. Show $|G| > 1$.

We define a new group from two old groups, which is more general than the direct product.

DEFINITION 6.4 (Semidirect product). Let N and H be groups and $\theta : H \to \mathrm{Aut}(N)$ be a homomorphism. The set $N \times H$ endowed with the multiplication

$$(n, h)(n_1, h_1) = (n\theta(h)(n_1), hh_1)$$

forms a group with identity (e_N, e_H) and $(n, h)^{-1} = (\theta(h^{-1})(n^{-1}), h^{-1})$, called the *semidirect product* of N and H, denoted by $N \times_\theta H$ or simply $N \rtimes H$.

REMARK 6.5. Let N and H be arbitrary groups and let $\theta : H \to \mathrm{Aut}(N)$ be the trivial map. Then the semidirect product is just the ordinary direct product.

Theorem 6.6. *Let G be a group. The following conditions are equivalent:*

(i) *There are subgroups N, H of G with N normal in G such that $NH = G$ and $N \cap H = 1$.*

(ii) $G \cong N \rtimes H$.

DEFINITION 6.7 (Wreath product). * Let G and H be finite groups with $H \leq S_n$. The *(permutation) wreath product* $G \wr H$ is the semidirect product of a normal subgroup N by H, where N is the direct product of n copies of G. The automorphism θ_h $(h \in H \leq S_n)$ of G^n is defined by $\theta_h(g_1, \ldots, g_n) = (g_{h(1)}, \ldots, g_{h(n)})$.

REMARK 6.8. * Since the map $H \to \mathrm{Aut}(G^n)$ by $h \mapsto \theta_h$ is a homomorphism, the semidirect product is well-defined. Note that $|G \wr H| = |G|^n|H|$.

EXAMPLE 6.9. * The group P_n of $n \times n$ permutation matrices, is isomorphic to the symmetric group S_n. Let B_n be the group of $n \times n$ matrices in which each nonzero entry of each matrix in P_n is allowed to be ± 1, so that B_n has a normal subgroup N of order 2^n consisting of diagonal matrices with each diagonal entry being either 1 or -1. Since each monomial matrix can be written as a product of a permutation matrix and a diagonal matrix, we obtain that $B_n = N \rtimes P_n$, i.e., $B_n \cong \mathbb{Z}/2\mathbb{Z} \wr S_n$. The case of each nonzero entry being allowed to be a power of $e^{2\pi i/r} \in \mathbb{C}$ is $\mathbb{Z}/r\mathbb{Z} \wr S_n$, which is called the *generalized symmetric group*, denoted by $G(r, 1, n)$.[10]

[10]The q-deformed group algebra of $G(r, 1, n)$ is called the *Ariki-Koike algebra*, which is a generalization of the *Hecke algebra*.

CHAPTER 4

Basic notions of rings

Dedekind[1] gave the first definitions of an axiomatic type of ring, ideal, field, and module in 1893. In his Lehrbuch der Algebra of 1894, Heinrich Weber[2] gave the first axiomatic definition of a group. The first thirty years of the 20th century, found several mathematicians, especially in Germany, setting up the framework for what would soon be called modern algebra. [D]

Ring theory comes from the study of polynomial rings and of algebraic number fields. Hilbert used the terminology *ring*[3], following the structure of $\mathbb{Z}/n\mathbb{Z}$. The set $\mathbb{Z}$ of integers with the usual addition and multiplication is the prototype of a ring.

1. Rings and fields

DEFINITION 1.1. A *ring* $(R, +, \cdot)$ is a set R with two binary operations, satisfying that

1) $(R, +)$ is an abelian group;
2) the operation $\cdot$ is associative;
3) the multiplication $\cdot$ is distributive with respect to the addition $+$, that is,[4]

$$a \cdot (b + c) = a \cdot b + a \cdot c \quad \text{and} \quad (a + b) \cdot c = a \cdot c + b \cdot c$$

for each $a, b, c \in R$.

A ring R is called *commutative* if $a \cdot b = b \cdot a$ for each $a, b \in R$.[5] The multiplicative identity (if it exists in a ring) is called the unity, usually denoted by 1.

[1]Julius Wilhelm Richard Dedekind(1831–1916), a German mathematician. He made important contributions to ring theory, algebraic number theory and the foundations of the real numbers.

[2]Heinrich Martin Weber(1842–1913), a German mathematician.

[3]We call it 환(環) in Korean.

[4]We say that two operations are compatible.

[5]The theory of commutative rings was given a firm axiomatic foundation by Emmy Noether(1882–1935, a German mathematician) in her monumental paper “Ideal theory in rings”(1921).

EXAMPLE 1.2. (1) The set $\mathbb{Z}$ of integers with addition and multiplication is a basic model of a ring. For a positive integer n, the set $n\mathbb{Z}$ of all multiples of n is a subring of $\mathbb{Z}$. The set

$$\mathbb{Z}_n = \{0, 1, 2, \ldots, n-1\}$$

with the modular addition and multiplication is also a ring.

(2) The set C of continuous real-valued functions of a real variable forms a ring, with addition and multiplication of functions.

(3) The set $M_{n \times n}(\mathbb{Q})$ of all $n \times n$ matrices with entries in $\mathbb{Q}$ is a ring under the matrix addition and multiplication. Note that $M_{n \times n}(\mathbb{Q})$ is not commutative if $n \geq 2$. What is the unity in $M_{n \times n}(\mathbb{Q})$? Find the multiplicative inverse of $\begin{pmatrix} 1 & 2 \\ 2 & 0 \end{pmatrix}$ in $M_{n \times n}(\mathbb{Z}_5)$?

(4) If R is a commutative ring with 1, so is the polynomial ring

$$R[x] := \{a_0 + a_1 x + \cdots + a_n x^n |\ a_i \in R, n \geq 0\}.$$

Find the additive inverse of $x^2 + x + 1$ in $\mathbb{Z}[x]$. Compute $(x+1)^5$ in $\mathbb{Z}_2[x]$.

Note that any polynomial $f(x) \in \mathbb{C}[x]$ has a root(or zero) α in $\mathbb{C}$, that is, there exists a solution $\alpha \in \mathbb{C}$ such that $f(\alpha) = 0$. This is said to be the *fundamental theorem of Algebra*, due to d'Alembert[6] in 1746 and Gauss in 1799, and so on.[7]

(5) Let $R_1, \ldots, R_n$ be rings. If we set

$$R = R_1 \times \cdots \times R_n = \{(a_1, \ldots, a_n)|\ a_i \in R_i\},$$

then R is a ring under the componentwise addition and multiplication.

PROPERTY 1.3. Let R be a ring and $a, b \in R$. Then

$$0a = a0 = 0, \quad a(-b) = (-a)b = -(ab), \quad (-a)(-b) = ab.$$

We define $a - b := a + (-b)$. Then

$$a(b-c) = ab - ac, \quad (a-b)c = ac - bc, \quad \text{and } (-1)a = -a, \quad (-1)(-1) = 1,$$

if R has 1.

EXERCISE 1.4. Prove the above equalities.

We define the notion of a morphism between objects.

[6]Jean-Baptiste le Rond d'Alembert(1717–1783), a French mathematician and music theorist.

[7]Various proofs are known. It is useful to see `http://navercast.naver.com/science/math/7884`.

DEFINITION 1.5. Let $(R,+,\cdot)$ and $(R',+',\cdot')$ be rings. A map $\phi : R \to R'$ is called a *homomorphism* of R into R' if it is an operation-preserving map, that is,

$$\phi(a+b) = \phi(a) +' \phi(b) \quad \text{and} \quad \phi(a \cdot b) = \phi(a) \cdot' \phi(b) \quad \text{for all } a, b \in R.$$

A bijective homomorphism is called an *isomorphism*. We say that two rings R and R' are *isomorphic*, denoted by $R \cong R'$, if there is an isomorphism of R onto R'.

EXAMPLE 1.6. The map $\phi : \mathbb{Z} \to \mathbb{Z}_n$, where $\phi(a)$ is the remainder of a modulo n is a ring homomorphism with $\ker \phi = n\mathbb{Z}$.

EXERCISE 1.7. (1) Find all ring homomorphisms of $\mathbb{Z}$ into $\mathbb{Z}$.

(2) Show that the rings $2\mathbb{Z}$ and $3\mathbb{Z}$ are not isomorphic.

DEFINITION 1.8. Let R be a ring with unity 1.

(a) An element $u \in R$ is called a *unit* if it has a multiplicative inverse v, denoted by u^{-1}, that is, there exists $v \in R$ such that $uv = vu = 1$

(b) If every nonzero element in R is a unit, then we say that R is a *division ring*. A commutative division ring is called a *field*.[8] A noncommutative division ring is called a skew-field.

REMARK 1.9. A unity in a ring R is unique if it exists. For $a \in R$, a multiplicative inverse of a is unique if it exists.

EXAMPLE 1.10. (1) $\mathbb{Z}$ is not a division ring. But, $\mathbb{Q}$, $\mathbb{R}$, $\mathbb{C}$ are fields.

(2) Note that $\mathbb{Z}_p$ is a field if and only if p is a prime.

EXERCISE 1.11 (Freshman exponentiation). Show that in the ring $\mathbb{Z}_p$ (p, a prime), we have

$$(a+b)^p = a^p + b^p \text{ for all } a, b \in \mathbb{Z}_p.$$

EXERCISE 1.12. (1) Which elements are units in $\mathbb{Z}$, $\mathbb{Q} \times \mathbb{Q}$, $M_{n\times n}(\mathbb{Q})$, $\mathbb{Q}[x]$, respectively?

(2) Let R be a ring with 1. Define $R^\times$ to be the set of all units in R. Show that $R^\times$ is a group under multiplication. Find $\mathbb{Z}_{20}^\times$.

[8]This name represents 운동장 or 마당, 즉 자유로운 공간, but we call it 체(體) meaning 몸, which originates from Körper in German. The finite field $\mathbb{F}_q$ with order $q = p^n$ for any prime p, plays an important role in the process of computers we are using everyday.

(3) Show that each of the following is a field:

$$\mathbb{Q}[\sqrt{2}] = \{a + b\sqrt{2} \mid a, b \in \mathbb{Q}\}, \quad \mathbb{Z}_7[i] = \{a + bi \mid a, b \in \mathbb{Z}_7\}.$$

(4) Prove that the fields $\mathbb{R}$ and $\mathbb{C}$ are not isomorphic.
[Hint: Any isomorphism $\phi : \mathbb{C} \to \mathbb{R}$ must satisfy that $\phi(1) = 1$ and $\phi(-1) = -1$.]

EXAMPLE 1.13. Consider the set

$$\mathbb{H} = \{a + bi + cj + dk \mid a, b, c, d \in \mathbb{R}\} = \{\alpha + \beta j \mid \alpha, \beta \in \mathbb{C}\}$$

of quaternions, equipped with the usual addition and the multiplication by

$$i^2 = j^2 = k^2 = ijk = -1.$$

We call it the *Hamiltonian skew-field.*[9] This is useful when we describe a rotation in the real 3-dimensional space.[10]

REMARK 1.14 (Frobenius 1877). *[11] We also remark that the division rings algebraic over $\mathbb{R}$ are only $\mathbb{R}$, $\mathbb{C}$ and $\mathbb{H}$.

Theorem 1.15 (Wedderburn[12] 1905). * *A finite division ring is a field.*

Proof. See [Hr, pp.361–367]. □

Theorem 1.16 (Category of rings). *

(a) *Let $\{R_i\}$ be a family of rings. Then $\prod R_i$ is a product in the category of rings.*
(b) *A commutative ring can be viewed as a commutative $\mathbb{Z}$-algebra. Then $\bigotimes R_i$ is a coproduct in the category of commutative rings.*
(c) *Let S be a nonempty set. A free object on S in the category of commutative rings is $\mathbb{Z}[S]$, the ring of polynomials with indeterminates in S over $\mathbb{Z}$.*

---

[9]This was discovered by William Rowan Hamilton(1805–1865), an Irish physicist and mathematician. On 16 October 1843, he was out walking along the Royal Canal in Dublin with his wife when the solution in the form of the equation $i^2 = j^2 = k^2 = ijk = -1$ suddenly occurred to him, promptly carving this equation using his penknife into the side of the nearby Brougham Bridge.

[10]For details, see `http://www.math.snu.ac.kr/~hongjong/잡필/QUATERNION.pdf`.

[11]For this theorem of Frobenius(1849–1917, a German mathematician), refer to [Hr, §7.3].

[12]Joseph Henry Maclagan Wedderburn(1882–1948), a Scottish mathematician. He taught at Princeton University for most of his career.

2. Subrings

We define the notion of a subobject.

DEFINITION 2.1. Let R be a ring, and $\emptyset \neq S \subset R$. We say that S is a *subring* of R if S is itself a ring under the operations in R. Denote by $S \leq R$.

Proposition 2.2. *Let R be a ring, and $\emptyset \neq S \subset R$. Then S is a subring of S if and only if $a - b, ab \in S$ for each $a, b \in S$.*

EXAMPLE 2.3. (1) $n\mathbb{Z} \leq \mathbb{Z}$ for $n \in \mathbb{Z}_{\geq 0}$. $\mathbb{Z} \leq \mathbb{Q}$.

(2) Let S be the set of polynomials in $\mathbb{Q}[x]$ whose value at 0 is 0. Then $S \leq \mathbb{Q}[x]$.

EXERCISE 2.4. (1) Let a be an element in a ring R. Let $S = \{x \in R |\ ax = 0\}$. Show that S is a subring.

(2) Let R be a commutative ring. Show that $S = \{r \in R |\ r^n = 0 \text{ for some } n \geq 1\}$ is a subring.

(3) The *center* of a ring R is the set $\{r \in R |\ ra = ar \text{ for all } a \in R\}$. Show that the center of R is a subring.

EXERCISE 2.5. (1) Define $\phi : \mathbb{C} \to M_{2\times 2}(\mathbb{R})$ by

$$\phi(a + bi) = \begin{pmatrix} a & b \\ -b & a \end{pmatrix}.$$

Show that ϕ is a ring homomorphism and injective, so

$$\mathbb{C} \cong \left\{ \begin{pmatrix} a & b \\ -b & a \end{pmatrix} \mid a, b \in \mathbb{R} \right\}$$

as a ring.

(2) Define $\phi : \mathbb{H} \to M_{2\times 2}(\mathbb{C})$ by

$$\phi(\alpha + \beta j) = \begin{pmatrix} \alpha & \beta \\ -\overline{\beta} & \overline{\alpha} \end{pmatrix}.$$

Show that ϕ is a ring homomorphism and injective. [Hint: Be cautious that $\beta j = j\overline{\beta}$.] Thus,

$$a + bi + cj + dk \overset{1-1}{\longleftrightarrow} a\begin{pmatrix} 1 & 0 \\ 0 & 1 \end{pmatrix} + b\begin{pmatrix} i & 0 \\ 0 & -i \end{pmatrix} + c\begin{pmatrix} 0 & 1 \\ -1 & 0 \end{pmatrix} + d\begin{pmatrix} 0 & i \\ i & 0 \end{pmatrix}.$$

EXERCISE 2.6. (1) An element a of a ring R is called *idempotent* if $a^2 = a$. If all elements in a ring R are idempotent, then show that $a + a = 0$ for all $a \in R$ and that

R is commutative. This ring R is called a *Boolean ring*,[13] whose explicit example appears in the next section.

(2) Show that the set of all idempotent elements of a commutative ring is closed under multiplication. Find all idempotents in $\mathbb{Z}_{20}$ and $\mathbb{Z}_6 \times \mathbb{Z}_{12}$. Show that the set of idempotent elements in $\mathbb{Z}_6$ under multiplication is isomorphic to the power set $P(\{1,2\})$ under intersection.

(3) Show that a division ring contains exactly two idempotent elements, and that $\mathbb{Z}_{p^n}$, where p is a prime and $n \in \mathbb{Z}_+$, also contains exactly two idempotents.

(4) Show that the number of idempotents in $\mathbb{Z}_n$ is 2^r, where r is the number of prime divisors of n.

Solution. (1) Let $a, b \in R$. From

$$a + a = (a+a)^2 = a^2 + a^2 + a^2 + a^2 = a + a + a + a,$$

we have $a + a = 0$ for each $a \in R$. On the other hand, the equality

$$a + b = (a+b)^2 = a^2 + ab + ba + b^2 = a + ab + ba + b$$

yields that $ab + ba = 0$. Together with $ab + ab = 0$, the commutativity follows. □

We list some more examples of rings.

EXAMPLE 2.7. (1) The set $M_{n\times n}(D)$ of all $n \times n$ matrices with entries in a division ring D is a noncommutative ring under the matrix addition and multiplication.

(2) *Polynomial rings*: $R[x], R[x_1, \dots, x_n]$ are integral domains, which is defined in the next section. The ring of multivariate polynomials is a principal object in computational algebra. In particular, $\mathbb{C}[x_1, \dots, x_n]$ is an object to be studied in algebraic geometry.

(3) *Subrings of* $\mathbb{C}$: $\mathbb{Z}[\alpha]$ denotes the smallest subring containing $\mathbb{Z}$ and $\alpha \in \mathbb{C}$. In particular,

$$\mathbb{Z}[i] = \{a + bi \mid a, b \in \mathbb{Z}\}$$

is called the ring of *Gaussian integers* and forms a square lattice in the complex plane. Another example is $\mathbb{Z}\left[\frac{1}{p}\right]$ consisting of rational numbers whose denominator is a power of p, which is a dense subset of $\mathbb{R}$.

[13]The term honors George Boole(1815–1864), an English mathematician.

(4) Any field is a ring. In particular, $\mathbb{Q}[\sqrt{d}] = \{a + b\sqrt{d} |\ a, b \in \mathbb{Q}\}$ where d is a square-free integer, is called a *quadratic number field.* In general, for $\alpha \in \mathbb{C}$ such that $f(\alpha) = 0$ for some $f \in \mathbb{Q}[x]$,

$$\mathbb{Q}[\alpha] = \{g(\alpha) |\ g \in \mathbb{Q}[x]\}$$

is called a *number field*, which is a main object in modern number theory.

3. Integral domains

The purposes in this section are:

1) Divide nonzero elements in $\mathbb{Z}_n$ into two classes;
2) Define an integral domain, an abstraction of $(\mathbb{Z}, +, \cdot)$;
3) Describe an inclusion relation among several types of rings; and finally,
4) Construct a field from an integral domain.

We focus on general facts concerning integral domains, whose important example is the ring of polynomials that will be treated thoroughly later.

DEFINITION 3.1. Let R be a ring. If two nonzero elements $a, b \in R$ satisfy $ab = 0$, then a and b are called *zero divisors.*

EXAMPLE 3.2. (1) Since $2 \cdot 3 = 0 \in \mathbb{Z}_6$, 2 and 3 are zero divisors.

(2) Find all solutions of the equation $x^2 + 3x + 2 = 0$ in $\mathbb{Z}_6$.

(3) Find all solutions of the equation $x^2 + x - 6 = 0$ in $\mathbb{Z}_{14}$.

(4) The matrix ring $M_{2\times 2}(\mathbb{Z}_2)$ has zero divisors: $\begin{pmatrix} 1 & 0 \\ 0 & 0 \end{pmatrix}\begin{pmatrix} 0 & 0 \\ 1 & 0 \end{pmatrix} = \begin{pmatrix} 0 & 0 \\ 0 & 0 \end{pmatrix}$.

(5) In the ring of 2×2 matrices over $\mathbb{R}$, $A^2 = I$ does not imply that $A = I$ or $A = -I$. A matrix $A = \begin{pmatrix} 1 & 0 \\ 1 & -1 \end{pmatrix}$ satisfies $A^2 = I$.

(6) A unit of a ring R is not a zero divisor.

Proposition 3.3. (a) *An element* $m \in \mathbb{Z}_n$ *is a zero divisor if and only if* $\gcd(m, n) > 1$.

(b) *An element* $m \in \mathbb{Z}_n$ *is a unit if and only if* $\gcd(m, n) = 1$.

Proof. (a) $\Leftarrow$) Let $d = \gcd(m, n) > 1$. Then $m\left(\frac{n}{d}\right) = \left(\frac{m}{d}\right)n = 0 \in \mathbb{Z}_n$.

$\Rightarrow$) Suppose that $\gcd(m, n) = 1$. If $ms = 0$ for some $s \in \mathbb{Z}_n$, then $n|ms$ implies that $n|s$, that is, $s = 0 \in \mathbb{Z}_n$.

(b) Note that $\gcd(m, n) = 1$ if and only if there exist s and t such that $ms + nt = 1$. □

Corollary 3.4. *The ring $\mathbb{Z}_p$ (p, a prime) has no zero divisors. In fact, it is a field.*

EXAMPLE 3.5. (1) $\mathbb{Z}_{12} = \{0\} \cup \{1, 5, 7, 11\} \cup \{2, 3, 4, 6, 8, 9, 10\}$. Here, $\{1, 5, 7, 11\}$ is the set of units, and $\{2, 3, 4, 6, 8, 9, 10\}$ is the set of zero divisors in $\mathbb{Z}_{12}$.

(2) In a ring R with 1, the set of units in R is a multiplicative group. In particular, $\mathbb{Z}_p^\times = \mathbb{Z}_p \setminus \{0\}$ is a group under multiplication. Moreover, we'll see that any finite subgroup of the multiplicative group of a field is cyclic. For example, $\mathbb{Z}_5^\times = \langle 2 \rangle$.

DEFINITION 3.6. Let R be a ring and $a, b, c \in R$. We say that the cancellation laws hold in R if $ab = ac$ with $a \neq 0$ implies $b = c$, and $ba = ca$ with $a \neq 0$ implies $b = c$.

Proposition 3.7. *The cancellation laws hold in a ring R if and only if R has no zero divisors.*

Now we define an algebraic object which does not have any zero divisor.

DEFINITION 3.8. A commutative ring with unity which has no zero divisors is called an *integral domain.*

EXAMPLE 3.9. $\mathbb{Z}$, $\mathbb{Z}_p$ (p, a prime), and $\mathbb{Z}[x]$ are integral domains. However, $\mathbb{Z}_n$ (n, a composite), $\mathbb{Z} \times \mathbb{Z}$ and $M_{2\times 2}(\mathbb{Z}_2)$ are not.

Theorem 3.10. (a) *Every field is an integral domain.*
(b) *Every finite integral domain is a field. So, $\mathbb{Z}_p$ (p, a prime) is a field.*

Proof. (a) Let $a \neq 0$ and b be elements in a field $\mathbb{F}$. If $ab = 0$, then the multiplication by a^{-1} from the left yields that $b = 0$. This means that every nonzero element in $\mathbb{F}$ is not a zero divisor.

(b) Let $D = \{0, a_0 = 1, a_1, \ldots, a_n\}$ be a finite integral domain. Take any nonzero $a \in D$. Then $aa_0, aa_1, aa_2 \ldots, aa_n$ are distinct by the cancellation law. Thus $aa_i = 1$ for some i, that is, any nonzero $a \in D$ has a multiplicative inverse. □

REMARK 3.11. Except for the finite field $\mathbb{Z}_p$ (p, a prime), what else are other things? That's an important question, which will be dealt in Ch.6 §3.

EXERCISE 3.12. Describe an inclusion relation among several types of rings: commutative rings, rings with 1, integral domains, division rings, fields, skew-fields.

EXERCISE 3.13. (1) Give an example of an integral domain which is not a field.

(2) Give an example of a commutative ring with 1 which is not an integral domain.

(3) Give an example of a commutative ring without 1.

(4) Five an example of a non-commutative ring with 1 which is not a skew-field.

(5) Five an example of a non-commutative ring without 1.

DEFINITION 3.14. Let R be a ring. For each $a \in R$, we consider n such that $n \cdot a = a + \cdots + a$ for n summands is equal to 0. The least $n \in \mathbb{Z}_+$ such that $n \cdot a = 0$ for all $a \in R$, is called the *characteristic* of R, denoted by $\operatorname{ch} R$. If no such n exists, we define to be $\operatorname{ch} R = 0$.

In case of $1 \in R$, $\operatorname{ch} R$ is the smallest $n \in \mathbb{Z}_+$ such that $n \cdot 1 = 0$ (if it exists).

EXAMPLE 3.15. $\operatorname{ch} \mathbb{Z}_n = n$, $\operatorname{ch} \mathbb{Z} = \operatorname{ch} \mathbb{Q} = \operatorname{ch} \mathbb{R} = \operatorname{ch} \mathbb{C} = 0$.

EXERCISE 3.16. Show that $\operatorname{ch}(\mathbb{Z}_n \times \mathbb{Z}_m) = \operatorname{lcm}(n, m)$.

Theorem 3.17. *The characteristic of an integral domain must be either* 0 *or a prime.*

Proof. Let D be an integral domain. Suppose that $\operatorname{ch} D = mn$ with $m \geq 2$ and $n \geq 2$, so $mn \cdot 1 = 0$ $m \geq 2$ and $n \geq 2$. Then $(m \cdot 1)(n \cdot 1) = 0 \in D$ implies that $m \cdot 1 = 0$ or $n \cdot 1 = 0$. This contradicts to the definition of $\operatorname{ch} D$. So, if $\operatorname{ch} D > 0$ then $\operatorname{ch} D$ must be a prime. □

EXERCISE 3.18. Let $S \neq \emptyset$ and $P(S)$ be the power set of S, i.e., the set of all subsets of S. For $A, B \in P(S)$, we define two binary operations to be

$$A + B := (A \cup B) \setminus (A \cap B), \quad A \cdot B := A \cap B.$$

Then $P(S)$ is a commutative ring with unity. Note that this is a *Boolean ring*,[14] which is isomorphic to $(\mathbb{Z}_2^{|S|}, +, \cdot)$.

(1) What is the additive identity and the unity in $P(S)$?

(2) Is $P(S)$ an integral domain? What is the characteristic of $P(S)$?

EXAMPLE 3.19. Consider $\prod \mathbb{Z}_p = \mathbb{Z}_2 \times \mathbb{Z}_3 \times \mathbb{Z}_5 \times \mathbb{Z}_7 \times \mathbb{Z}_{11} \times \cdots$, where p is over all primes, with the componentwise addition and multiplication. Then this is a commutative ring with unity which is not an integral domain, with characteristic 0.

[14]See Exercise 2.6. A Boolean algebra plays an important role in discrete mathematics, etc. It is used for electric circuits [G, Ch.34]. By Stone's representation theorem(Stone 1973), every finite Boolean algebra is an algebra of sets [유, Ch.7].

EXERCISE 3.20. Let R be a commutative ring with unity of prime characteristic p. Show that the map $f_p : R \to R$ given by $f_p(x) = x^p$ is a homomorphism. We call f_p the *Frobenius homomorphism.*

EXERCISE 3.21. Let $\mathbb{F}$ be a field of characteristic zero and let D be the formal polynomial differentiation map, so that

$$D(a_0 + a_1x + a_2x^2 + \cdots + a_nx^n) = a_1 + 2a_2x + \cdots + na_nx^{n-1}.$$

(1) Show that $D : \mathbb{F}[x] \to \mathbb{F}[x]$ is a additive group homomorphism. Is D a ring homomorphism?

(2) Find the kernel of D.

(3) Find the image of $\mathbb{F}[x]$ under D.

Now we consider how to construct a field from a given integral domain, which is a generalization of the method to construct $\mathbb{Q}$ from $\mathbb{Z}$.

Theorem 3.22. *Let D be an integral domain. Then there exists a field F containing D.*

Proof. Let $S = \{(a,b)|\ a,b \in D, b \neq 0\}$, and define a relation $\sim$ on S by

$$(a,b) \sim (c,d) \text{ if and only if } ad = bc.$$

Then $\sim$ is an equivalence relation, so we can consider the equivalence class $[a,b]$ of (a,b). We set

$$F = \{[a,b]|\ a,b \in D, b \neq 0\}$$

and define the binary operations on F to be

$$[a,b] + [c,d] := [ad + bc, bd], \quad [a,b] \cdot [c,d] := [ac, bd].$$

Now we have to check that these operations are well-defined. For $[a,b] = [a',b']$ and $[c,d] = [c',d']$, that is, $ab' = a'b$ and $cd' = c'd$, we check that

$$(ad + bc)b'd' - (a'd' + b'c')bd = (ab' - a'b)dd' + bb'(cd' - c'd) = 0,$$

that is, $[ad + bc, bd] = [a'd' + b'c', b'd']$. It can be checked that the multiplication is also well-defined. We can check that F is a commutative ring with unity $[1,1]$. Furthermore, if $[a,b] \neq [0,1]$, then $[a,b][b,a] = [1,1]$. Thus, F is a field.

Define a map $\phi : D \to F$ by $\phi(a) = [a, 1]$. Then ϕ is a ring homomorphism with $\ker \phi = \{0\}$. This means that D is isomorphic to $\operatorname{Im} \phi$, a subring of F. $\square$

Note that F is the smallest field containing D. We call this F the *quotient field* of D, or the *field of fractions* of D.

EXAMPLE 3.23. (1) The quotient field of $\mathbb{Z}$ is isomorphic to $\mathbb{Q}$. For $\mathbb{Z}[\sqrt{2}]$, its quotient field is isomorphic to $\mathbb{Q}[\sqrt{2}]$. The quotient field of the domain $\mathbb{Z}[i]$ of Gaussian integers is $\mathbb{Q}[i]$.

(2) Let D be an integral domain. Then $D[x]$ is also an integral domain. We define

$$D(x) = \left\{ \frac{f(x)}{g(x)} \mid f, g \in D[x], g \neq 0 \right\},$$

the set of all rational polynomials. Then the quotient field of $D[x]$ is isomorphic to $D(x)$ under the map $[f(x), g(x)] \mapsto \frac{f(x)}{g(x)}$.

4. Applications of $\mathbb{Z}_n$

We deal with some famous theorems in number theory, concerning the structure of the ring $\mathbb{Z}_n$.

Let us understand the structure of $\mathbb{Z}/n\mathbb{Z}$ with the well-defined addition and multiplication. Then the following theorems, named after Fermat and Euler, are easy to understand.

Theorem 4.1 (Fermat's little theorem[15]). *Let p be a prime and $a \in \mathbb{Z}$ with $p \nmid a$. Then $a^{p-1} - 1$ is divisible by p, that is,*

$$a^{p-1} \equiv 1 \pmod{p}.$$

Proof. Consider $\mathbb{Z}_p^\times = \{1, 2, \ldots, p-1\}$, which is a group under the modular multiplication. So, for each $b \in \mathbb{Z}_p^\times$, we have $b^{p-1} = 1 \in \mathbb{Z}_p^\times$. In other words, since $\mathbb{Z}_p \cong \mathbb{Z}/p\mathbb{Z}$, any element $a \in \mathbb{Z}$ with $a \notin 0 + p\mathbb{Z}$ satisfies $(a + p\mathbb{Z})^{p-1} = a^{p-1} + p\mathbb{Z} = 1 + p\mathbb{Z}$. $\square$

EXAMPLE 4.2. (1) For each $a \in \mathbb{Z}$, $a^p \equiv a \pmod{p}$, where p is a prime.

(2) $3^{10} \equiv 1 \pmod{11}$, $89^{820} \equiv 1 \pmod{821}$, $2^{1137} = (2^{16})^{71} \cdot 2 \equiv 2 \pmod{17}$.

(3) $2^{11213} - 1$ is not divisible by 11, because $2^{10} \equiv 1 \pmod{11}$ leads to $2^{11213} - 1 = (2^{10})^{1121} \cdot 2^3 - 1 \equiv 2^3 - 1 = 7 \pmod{11}$.

---

[15]Pierre de Fermat(1601–1665), a French lawyer and an amateur mathematician.

DEFINITION 4.3. A prime number of the form $2^p - 1$ is called a *Mersenne prime.*[16]

EXAMPLE 4.4. The greatest Mersenne prime ever found up to the present[17] is $2^{57885161} - 1$ (over 17.4 million digits).[18]

EXERCISE 4.5. Use Fermat's little theorem to find all zeros in $\mathbb{Z}_5$ of $2x^{219} + 3x^{74} + 2x^{57} + 3x^{44}$.

We generalize Fermat's little theorem.

Proposition 4.6. *Let $\mathbb{Z}_n^\times$ be the set of nonzero elements in $\mathbb{Z}_n^\times$ which are not zero divisors. Then $\mathbb{Z}_n^\times$ is a group under the multiplication modulo n.*

Proof. First, we claim that $\mathbb{Z}_n^\times$ is closed under the multiplication modulo n. Suppose that $a, b \in \mathbb{Z}_n^\times$ but $ab \notin \mathbb{Z}_n^\times$. Then there exists nonzero $c \in \mathbb{Z}_n$ such that $(ab)c = 0$. From $a(bc) = 0 \in \mathbb{Z}_n$, we have $bc = 0 \in \mathbb{Z}_n$, contradiction to that $b \in \mathbb{Z}_n^\times$.

Clearly, the multiplication in $\mathbb{Z}_n^\times$ is associative and $1 \in \mathbb{Z}_n^\times$. We claim that every element in $\mathbb{Z}_n^\times$ has a multiplicative inverse. Let $\mathbb{Z}_n^\times = \{1 = a_0, a_1, \ldots, a_r\}$ and $a \in \mathbb{Z}_n^\times$. Then the elements $aa_0, aa_1, \ldots, aa_r$ are distinct in $\mathbb{Z}_n^\times$ by the cancellation law coming from the fact that a is not a zero divisor. Thus, $aa_i = 1$ for some i. We proved. □

DEFINITION 4.7. Let n be a positive integer. The value $\phi(n)$ is defined to be the number of positive integers less than n which are relatively prime to n. The function $\phi : \mathbb{Z}_+ \to \mathbb{Z}_+$ is called the *Euler phi function.*

EXAMPLE 4.8. $\phi(90) = \phi(2 \cdot 3^2 \cdot 5) = \phi(2)\phi(3^2)\phi(5) = (2-1)(9-3)(5-1) = 24$.

Theorem 4.9 (Euler). *Let a and n be positive integers which are relatively prime. Then $a^{\phi(n)} - 1$ is divisible by n, i.e.,*

$$a^{\phi(n)} \equiv 1 \pmod{n}.$$

[16] Marin Mersenne(1588–1648), a French theologian, mathematician and music theorist.

[17] February 2014. Refer to the website `http://primes.utm.edu/largest.html`. The security of RSA cryptosystem is based on the fact that it it very hard to factorize a large integer into a product of primes. So, finding a large prime is practically important.

[18] A positive integer n is a perfect number if n is the sum of its positive divisors less than n. The integers $6, 28, 486 = 2^4(2^5-1), 8128 = 2^6(2^7-1)$ are perfect. Note that not all numbers of the form $2^p - 1$ (p, a prime) are prime. e.g.) $2^{11} - 1 = 2047 = 23 \times 89$. It is known that finding an even perfect number is equivalent to finding a prime of the form $2^p - 1$. It is unknown whether there is any odd perfect number. [오이, §7.4]

Proof. Consider the multiplicative group $\mathbb{Z}_n^\times$ in $\mathbb{Z}_n$, whose order is $\phi(n)$. So, for each $a \in \mathbb{Z}_n^\times$, we have $a^{\phi(n)} = 1 \in \mathbb{Z}_n^\times$. □

EXERCISE 4.10. (1) Find the remainder of 7^{100} when divided by 24. [Hint: Euler's theorem]

(2) Find the remainder of 11^{1000} when divided by 100.

(3) Compute the remainder of $2^{(2^{17})}+1$ when divided by 19. [Hint: You will need to compute the remainder of 2^{17} modulo 18. Use $\phi(18) = 6$.]

We apply Euler's theorem to solving linear congruences.

Theorem 4.11. *Let n be a positive integer and $a \in \mathbb{Z}_n$ with $\gcd(a, n) = 1$. Then, for each $b \in \mathbb{Z}_n$, a linear equation $ax = b$ has a unique solution in $\mathbb{Z}_n$. In other words, a linear congruence*

$$ax \equiv b \pmod{n}$$

has, as solutions, all integers in precise one residue class modulo n.

Proof. The condition $\gcd(a, n) = 1$ means that a is a unit in $\mathbb{Z}_n$. So $ax = b$ has a unique solution $x = a^{-1}b \in \mathbb{Z}_n$. □

EXAMPLE 4.12. Solve $5x \equiv 3 \pmod{6}$. Note that $5^{-1} = 5 \in \mathbb{Z}_6$. So this congruence has a unique solution $x = 5^{-1} \cdot 3 = 5 \cdot 3 = 3 \in \mathbb{Z}_6$.

Theorem 4.13. *Let n be a positive integer and $a, b \in \mathbb{Z}_n$ with $d := \gcd(a, n) \geq 1$. Then a linear equation $ax = b$ has a solution in $\mathbb{Z}_n$ if and only if b is divisible by d. In this case, the number of solutions in $\mathbb{Z}_n$ is d.*

Proof. $\Rightarrow$) If there is a solution s such that $as = b$, then clearly $a|b$, so b is a multiple of d.

$\Leftarrow$) Let $a = a_1 d$ and $b = b_1 d$ and $n = n_1 d$, where $\gcd(a_1, n_1) = 1$. Since $a_1 x = b_1$ has a unique solution $s = a_1^{-1} b_1 \in \mathbb{Z}_{n_1}$, the congruence $d(a_1 x - b_1) \equiv 0 \pmod{dn_1}$ has solutions $s, s + n_1, s + 2n_1, \ldots, s + (d-1)n_1 \in \mathbb{Z}_n$. □

EXAMPLE 4.14. (1) A congruence $12x \equiv 27 \pmod{18}$ has no solution, because $\gcd(12, 18) = 6$ does not divide 27.

(2) Let us solve $15x \equiv 9 \pmod{18}$. Divide by $\gcd(15, 18) = 3$ to get a reduced congruence $5x \equiv 3 \pmod{6}$. This has a unique solution $x = 5^{-1} \cdot 3 = 5 \cdot 3 = 3 \in \mathbb{Z}_6$. So the original congruence has solutions $3, 3 + 6 = 9, 3 + 12 = 15$ in $\mathbb{Z}_{18}$.

EXERCISE 4.15. (1) Describe all solutions of $45x \equiv 15 \pmod{24}$.

(2) Describe all solutions of $17x \equiv 3 \pmod{90}$.

REMARK 4.16. For given two large integers, it is not easy to find the gcd. An efficient method is *Euclidean algorithm*, which is described at the end of this section, with *extended Euclidean algorithm*.

We state and prove another useful theorem.

Theorem 4.17 (Wilson[19]). *Let p be a prime. Then*

$$(p-1)! \equiv -1 \pmod{p}.$$

Proof. If $p = 2, 3$ then the theorem holds. Let $p > 3$. Each $x \in \mathbb{Z}_p^\times = \{1, 2, \ldots, p-1\}$ has the inverse $x^{-1} \in \mathbb{Z}^\times$. We solve $x = x^{-1}$, that is, $x^2 = 1 \in \mathbb{Z}_p^\times$. From $x^2 - 1 = (x-1)(x+1) \equiv 0 \pmod{p}$, we get $x = 1, p-1 \in \mathbb{Z}_p^\times$. So $(p-2)! = r_1 r_1^{-1} r_2 r_2^{-1} \cdots r_s r_s^{-1} \equiv 1 \pmod{p}$ for $s = \frac{p-3}{2}$. Thus, $(p-1)! \equiv p - 1 \equiv -1 \pmod{p}$. □

EXAMPLE 4.18. If $p = 7$ then $6! = (2 \cdot 4) \cdot (3 \cdot 5) \cdot 6 \equiv -1 \pmod{7}$. Another example is $10! = (2 \cdot 6) \cdot (3 \cdot 4) \cdot (5 \cdot 9) \cdot (7 \cdot 8) \cdot 10 \equiv -1 \pmod{11}$.

REMARK 4.19. (1)(The converse of Wilson's theorem) If $(n-1)! \equiv -1 \pmod{n}$ for a positive integer n, then n is a prime.

(2) Wilson's theorem implies that composite numbers of the form $n! + 1$ are infinitely many. But, it is not known how many prime numbers of the form $n! + 1$ are there.

The primes of the form $n!+1$ when $n \leq 100$ are the cases of $n = 1, 2, 3, 11, 27, 37, 41, 73, 77$. The integer $26951! + 1$ is a prime.[20]

In the following theorem, Fermat's little theorem and Wilson's theorem play a crucial role.

Theorem 4.20. *Let p be a prime ≥ 3. Then $x^2 \equiv -1 \pmod{p}$ has a solution if and only if $p \equiv 1 \pmod{4}$. In this case, $x \equiv \left(\frac{p-1}{2}\right)!$.*

---

[19] Edward Waring(1736–1798, an English mathematician) announced the theorem in 1770, although neither he nor his student John Wilson(1741–1793) could prove it. Lagrange gave the first proof in 1771. In fact, Leibniz(Gottfried Wilhelm von Leibniz, 1646–1716, a German mathematician and philosopher) was also aware of the result a century earlier, but he never published it.

[20] For the factorial primes ever known, see `http://primes.utm.edu/top20/page.php?id=30`.

Proof. $\Rightarrow$) Combined with $x^{p-1} \equiv 1 \pmod{p}$, we get that $x^{p-1} = x^{2\cdot\frac{p-1}{2}} \equiv 1 \pmod{p}$, so we deduce that $\frac{p-1}{2}$ must be even.

$\Leftarrow$) We calculate that

$$\begin{aligned}
(p-1)! &= 1\cdot 2\cdots \tfrac{p-1}{2}\cdot (p-1)(p-2)\cdots\left(p-\tfrac{p-1}{2}\right)\\
&\equiv 1\cdot 2\cdots \tfrac{p-1}{2}\cdot (-1)\cdot(-2)\cdots\left(-\tfrac{p-1}{2}\right)\\
&= (-1)^{\frac{p-1}{2}}1^2\cdot 2^2\cdots\left(\tfrac{p-1}{2}\right)^2 = \left(1\cdot 2\cdots\left(\tfrac{p-1}{2}\right)\right)^2 \pmod{p}.
\end{aligned}$$

Together with $(p-1)! \equiv -1 \pmod{p}$, $x = \left(\frac{p-1}{2}\right)!$ is a solution of $x^2 \equiv -1 \pmod{p}$. □

EXERCISE 4.21. Solve the congruences $x^2 \equiv -1 \pmod{17}$ and $x^2 \equiv -1 \pmod{43}$.

EXERCISE 4.22. State and prove the following briefly:
(1) Fermat's little theorem, (2) Euler's theorem, and (3) Wilson's theorem.

4.1. Euclidean algorithm.

Lemma 4.23. *Let a and b be any nonzero distinct integers. Then*

$$\gcd(a,b) = \gcd(a, b-a).$$

Proof. Take any d such that $d|a$ and $d|b$. Then $d|b-a$. So $\gcd(a,b) \le \gcd(a,b-a)$.
We replace a, b by $-a$, $b-a$, respectively. Then

$$\gcd(a,b-a) = \gcd(-a,b-a) \le \gcd(-a,b) = \gcd(a,b),$$

which proves the identity. □

We present a crucial algorithm, due to Euclid[21].

Theorem 4.24 (Division algorithm). *For $a, b \in \mathbb{Z}$ with $b > 0$, there exist a unique quotient $q \in \mathbb{Z}$ and the remainder $r \in \mathbb{Z}$ such that*

$$a = qb + r \quad (0 \le r < b).$$

[21]Euclid(BC 325–265, Alexsandria) wrote Elements, 13 books of geometric and other mathematical treatise.

Proof. In the enumeration of integers

$$\ldots,\ a-2b,\ a-b,\ a,\ a+b,\ a+2b,\ \ldots,$$

there exists the least nonnegative element, say r, i.e., $0 \le r = a - qb$. Suppose that $r \ge a$. Then $0 \le r - b = a - (q+1)b$, which is a contradiction. The uniqueness follows because we let $a = qb + r$ $(0 \le r < b)$ and $a = q_1 b + r_1$ $(0 \le r_1 < b)$ implying that $-b < r - r_1 = (q - q_1)b < b$ thus $r - r_1 = 0$. □

EXERCISE 4.25. (1) Find the quotient and remainder when we divide -7 by 3.

(2) Find the quotient and remainder when we divide 710 by 68.

REMARK 4.26. An integral domain on which the division algorithm works is called a *Euclidean domain*. The ring $\mathbb{Z}$ of integers, the ring $\mathbb{F}[x]$ of polynomials in one variable over a field $\mathbb{F}$, and the domain $\mathbb{Z}[i]$ of Gaussian integers are typical examples of Euclidean domains.

We know that the multiplication of two integers is easy, however, the factorization of a large integer into primes is hard. So, computing the gcd by factorization is not easy.

Now we present an efficient method to find the gcd of integers. How to find the gcd of two large integers? For example, how to find quickly the integers

$$\gcd(2013, 291)\ ?$$

EXERCISE 4.27. Find the following numbers:

$$\gcd(2010, 737),\ \gcd(2009, 943),\ \gcd(2008, 1757),\ \gcd(2007, 1561),\ \gcd(2006, 901).$$ [22]

Is it easy to factorize the integer 2006 into primes?

Following Lemma 4.23, we have an algorithm to find the gcd without factorization.

[22] The answers are $67, 41, 251, 223$ and 17, respectively. The number 2011 is a prime, and $2012 = 2^2 \cdot 503$, $2013 = 3 \cdot 11 \cdot 61$, $2014 = 2 \cdot 19 \cdot 53$.

Theorem 4.28 (Euclidean algorithm). *For two positive integers $a(=r_0), b(=r_1)$ with $a > b$,*

$$\begin{aligned} a &= q_0 b + r_2 \quad (0 < r_2 < b), \\ b &= q_1 r_2 + r_3 \quad (0 < r_3 < r_2), \\ r_2 &= q_2 r_3 + r_4 \quad (0 < r_4 < r_3), \\ &\vdots \\ r_{m-1} &= q_{m-1} r_m + r_{m+1} \quad (0 < r_{m+1} < r_m), \\ r_m &= q_m r_{m+1}. \end{aligned}$$

Then we conclude that

$$\gcd(a, b) = \gcd(b, r_2) = \gcd(r_2, r_3) = \cdots = \gcd(r_m, r_{m+1}) = r_{m+1}.$$

Exercise 4.29. Find $\gcd(710, 68)$.

Moreover, what are $s, t \in \mathbb{Z}$ such that $as + bt = \gcd(a, b)$?

Theorem 4.30 (Extended Euclidean algorithm). *With the same notations as in the previous theorem, we let*

$$s_0 = 1, s_1 = 0, s_i = s_{i-2} - q_{i-2}s_{i-1} \quad \textit{and} \quad t_0 = 0, t_1 = 1, t_i = t_{i-2} - q_{i-2}t_{i-1}$$

for $2 \leq i \leq m+1$. Then it holds that

$$as_i + bt_i = r_i$$

for $0 \leq i \leq m+1$. In particular,

$$as_{m+1} + bt_{m+1} = \gcd(a, b).$$

Proof. Use induction on i. Clearly, the cases of $i = 0$ and $i = 1$ are okay.

Assume that the formula holds for $i = n-2$ and $i = n-1$ $(n \geq 2)$. Then

$$\begin{aligned} as_n + bt_n &= a(s_{n-2} - q_{n-2}s_{n-1}) + b(t_{n-2} - q_{n-2}t_{n-1}) \\ &= as_{n-2} + bt_{n-2} - q_{n-2}(as_{n-1} + bt_{n-1}) \\ &= r_{n-2} - q_{n-2}r_{n-1} = r_n, \end{aligned}$$

which is the case of $i = n$.

Thus we proved for $i = 0, 1, 2, \ldots, m+1$. □

Example 4.31. Find integers s, t such that $710s + 68t = \gcd(710, 68)$. We calculate that

q_i	10	2	3	1		
s_i	1	0	1	−2	7	−9 .
t_i	0	1	−10	21	−73	94

Thus one possible solution is $s = -9$ and $t = 94$. Note that every solution is of the form $s = -9 + 68k$, $t = 94 - 710k$ ($k \in \mathbb{Z}$).

EXERCISE 4.32. (1) Show that 71 and 45 are relatively prime. Find integers s and t such that $71s + 45t = 1$. [23]

(2) Show that $\gcd(2261, 1275) = 17$. What are integers s and t such that $2261s + 1275t = 17$? [24]

REMARK 4.33. * (1) Calculating a greatest common divisor is an essential step in several integer factorization algorithms. See [CP].

(2) The complexity for computing $\gcd(a, b)$ with $a > b > 0$ is, to mention briefly, proportional to the number of digits of b. In 1841, P.-J.-E. Finck showed that the number of division steps is at most $2\log_2 b + 1$, and hence Euclid's algorithm runs in time polynomial in the size of the input, which is very efficient. Gabriel Lamé in 1844 showed that the number of steps required for completion is never more than five times the number of base-10 digits of the smaller number.

REMARK 4.34. (1) For a real number, we divide it into its integer part and decimal part. The decimal is taken to be its inverse, so we divide again. This process is called the *continued fraction*. For example, a rational number can be written as a finite continued fraction by applying the division algorithms repeatedly:

$$\frac{710}{68} = 10 + \frac{1}{\frac{68}{30}} = 10 + \frac{1}{2 + \frac{1}{\frac{30}{8}}} = 10 + \frac{1}{2 + \frac{1}{3 + \frac{1}{\frac{8}{6}}}} == 10 + \frac{1}{2 + \frac{1}{3 + \frac{1}{1 + \frac{1}{3}}}},$$

So we write

$$\frac{710}{68} = [10; 2, 3, 1, 3].$$

We note that a real number is rational if and only if it has a finite continued fraction.

(2) The division algorithm and the Euclidean algorithm are applied to the case of univariate polynomials. For instance, we consider

$$f_1 = x^4 - 2x^3 - x^2 + 2x, \quad f_2 = x^3 - 7x^2 + 14x - 8$$

[23] One of the possible answers is $s = -19, t = 30$.

[24] One solution is $(s, t) = (22, -39)$.

with rational coefficients. Then f_1 is divided by f_2:

$$f_1 = (x+5)f_2 + x^2 - 3x + 2.$$

So the greatest common divisor is

$$\gcd(f_1, f_2) = \gcd(f_2, x^2 - 3x + 2) = x^2 - 3x + 2.$$

5. Rings of polynomials

In this section, we introduce the ring $\mathbb{F}[x]$ of polynomials over a field $\mathbb{F}$. Our focus is to understand the similarity of structures between $\mathbb{Z}$ and $\mathbb{F}[x]$.

DEFINITION 5.1. Let R be a ring. Then a finite sum of monomials

$$a_0 + a_1x + a_2x^2 + \cdots + a_nx^n,$$

where $a_i \in R$ ($a_n \neq 0$) and x is an indeterminate(or variable), is called a *polynomial* with coefficients in R. The nonnegative integer n is the *degree* of $f(x)$. (By convention, $\deg 0 = -\infty$.) We call a_n the *leading coefficient* of $f(x)$, denoted by $\mathrm{lc}(f)$. A polynomial $f(x)$ with $\mathrm{lc}(f) = 1$ is called *monic*. We identify $f(x)$ with the ordered set of its coefficients $(a_0, a_1, \ldots, a_n)$.

Theorem 5.2. (a) *We define the addition and multiplication of two polynomials as usual: For $f(x) = \sum_{i=0}^{\infty} a_ix^i$ and $g(x) = \sum_{i=0}^{\infty} b_ix^i$, where all but finitely many a_i's and b_i's are zeros,*

$$f(x) + g(x) := \sum_{i=0}^{\infty}(a_i + b_i)x^i, \qquad f(x)g(x) := \sum_{i=0}^{\infty}\left(\sum_{j=0}^{i} a_jb_{i-j}\right)x^i.$$

Then the set $R[x]$ of all polynomials with coefficients in R is a ring. Moreover, if R is commutative then $R[x]$ is commutative. If $1 \in R$ then $1 \in R[x]$.

(b) *If D is an integral domain, then $D[x]$ is also an integral domain. In particular, if $\mathbb{F}$ is a field, then $\mathbb{F}[x]$ is an integral domain. Note that $\mathbb{F}[x]$ is not a field, since x is not a unit.*

Proof. (b) Let $f(x)$ and $g(x)$ be nonzero polynomials in $D[x]$, where $\deg f(x) = m \geq 0$, $\deg g(x) = n \geq 0$, and $\mathrm{lc}(f) = a_m$, $\mathrm{lc}(g) = b_n$. Then $\deg f(x)g(x) = m + n$ and $\mathrm{lc}(fg) = a_mb_n \neq 0$, thus $f(x)g(x) \neq 0$. □

EXERCISE 5.3. Let D be an integral domain and $f(x)$, $g(x)$ non-zero polynomials in $D[x]$. Show that $\deg\left(f(x)g(x)\right) = \deg f(x) + \deg g(x)$.

Definition 5.4. Let $F \leq E$ be fields and $\alpha \in E$. Then the map $\phi_\alpha : F[x] \to E$ defined by

$$a_0 + a_1x + \cdots + a_nx^n \mapsto a_0 + a_1\alpha + + \cdots + a_n\alpha^n$$

is a homomorphism. (Check this.) In fact, for $F \leq E$ be commutative rings with 1, this homomorphism is defined. This map ϕ_α is called the *evaluation homomorphism at* α. Note that $\phi_\alpha(x) = \alpha$ and $\phi_\alpha(a_0) = a_0$ for $a_0 \in F$.

Example 5.5. (1) Consider $\mathbb{Q} < \mathbb{R}$ and the evaluation homomorphism $\phi_\alpha : \mathbb{Q}[x] \to \mathbb{R}$. Then ϕ_0 is the map sending a polynomial $f(x) = a_0 + a_1x + a_2x^2 + \cdots + a_nx^n$ to its constant term a_0. In addition, $\phi_2(x-2) = 0$ and $\phi_{\sqrt{2}}(x^2-2) = 0$.

(2) However, $\phi_\pi(f(x)) \neq 0$ for any polynomial $f(x) \in \mathbb{Q}[x]$. That is, $\pi \in \mathbb{R}$ is not an algebraic number,[25] and ϕ_π is an isomorphism from $\mathbb{Q}[x]$ onto $\mathbb{Q}[\pi] < \mathbb{R}$.

(3) If we consider $\phi_\alpha : \mathbb{Q}[x] \to \mathbb{C}$, we get that $\phi_i(x^2+1) = 0$. Note that $i \in \mathbb{C} \setminus \mathbb{R}$. It is important in which field the element α of ϕ_α lies.

Definition 5.6. Let $F \leq E$ be fields, $f(x) \in F[x]$. Then $\alpha \in E$ is called a *zero*(or *root*) of $f(x)$ if $f(\alpha) = \phi_\alpha(f(x)) = 0$.

Our basic goal is: For a field F and a polynomial $f(x) \in F[x]$, we want to construct an extension field E of F in which $f(x)$ has a zero. We will study this topic in Chapter 6.

We deal with factorization of polynomials. The following theorem is one of the most fundamental properties in algebra, leading to the extended Euclidean algorithm for finding the greatest common divisor.

Theorem 5.7 (Division algorithm for polynomials over a field). *Let $\mathbb{F}$ be a field, and $f(x), g(x) \in \mathbb{F}[x]$ with $\deg f(x) \geq \deg g(x) \geq 1$. Then there exist unique polynomials $q(x)$ and $r(x)$ in $\mathbb{F}[x]$ such that*

$$f(x) = g(x)q(x) + r(x), \text{ and } \deg r(x) < \deg g(x).$$

We call the $q(x)$ and $r(x)$ the *quotient* and *remainder*, respectively, in the division of $f(x)$ by $g(x)$.

[25] This was proved by Lindemann in 1882, using some facts on functions of a complex variable [L, Appendix 1]. For an overview to transcendental numbers, see [김김, HP7].

Proof. First, we prove the existence of $q(x)$ and $r(x)$. Consider the set

$$S = \{f(x) - g(x)p(x) \mid p(x) \in \mathbb{F}[x]\}.$$

If $0 \in S$ then there exists $p(x)$ such that $f(x) = g(x)p(x)$, so we take $q(x) = p(x)$ and $r(x) = 0$. If $0 \notin S$, then let $r(x)$ be an element of minimal degree in S. Thus, $f(x) = g(x)q(x) + r(x)$ for some $q(x) \in \mathbb{F}[x]$. We have only to show that $\deg r(x) < \deg g(x)$.

Suppose that $\deg r(x) \geq \deg g(x)$. Let $g(x) = b_n x^n + \cdots + b_0$ ($b_n \neq 0$) and $r(x) = c_t x^t + \cdots + c_0$ ($c_t \neq 0$). Then $f(x) - g(x)q(x) - (c_t/b_n)x^{t-n}g(x)$ is a polynomial of degree less than t, which belongs to S. This contradicts to the minimality of $\deg r(x)$. Hence, we proved that $\deg r(x) < \deg g(x)$.

The uniqueness follows easily. □

EXERCISE 5.8. Compute the quotient and remainder when $f(x)$ is divided by $g(x)$.

(1) $f(x) = x^3 + 2x + 4$, $g(x) = 3x + 2$ in $\mathbb{Z}_5[x]$.

(2) $f(x) = 3x^4 + x^3 + 2x^2 + 1$, $g(x) = x^2 + 4x + 2$ in $\mathbb{Z}_5[x]$.

(3) $f(x) = 5x^4 + 3x^3 + 1$, $g(x) = 3x^2 + 2x - 1$ in $\mathbb{Z}_7[x]$.

Corollary 5.9. *Let $\mathbb{F}$ be a field and $f(x) \in \mathbb{F}[x]$.*

(a) *An element $a \in \mathbb{F}$ is a zero of $f(x)$ if and only if $x - a$ is a factor of $f(x)$ over $\mathbb{F}$.*

(b) *If $\deg f(x) = n$, then the number of zeros of $f(x)$ in $\mathbb{F}$ is at most n.*

REMARK 5.10. A polynomial in $R[x]$ (R, a ring) of degree m may have zeros more than m. e.g.) $x^2 - 1 \in \mathbb{Z}_{12}$ has 4 zeros in $\mathbb{Z}_{12}$. Note that $\mathbb{Z}_{12}$ is not an integral domain.

EXERCISE 5.11. (1) Let p be a prime. Show that

$$x^{p-1} - 1 = (x-1)(x-2)\cdots(x-p+1) \in \mathbb{Z}_p[x].$$

(2) Let $\mathbb{F}$ be an infinite field and $f(x) \in \mathbb{F}[x]$. Show that if $f(a) = 0$ for infinitely many $a \in \mathbb{F}$, then $f(x) = 0$.

EXERCISE 5.12. Let $\mathbb{F}$ be a field of characteristic 0. For $f(x) = \sum_{k=0}^{n} a_k x^k \in \mathbb{F}[x]$, we define the *formal derivative* of $f(x)$ by

$$f'(x) = \sum_{k=1}^{n} k a_k x^{k-1}.$$

(1) If $f(a) = 0$ but $f'(a) \neq 0$ for some $a \in \mathbb{F}$, then show that the multiplicity of a in $f(x)$ is 1, that is, show that

$$x - a | f(x) \text{ but } (x-a)^2 \nmid f(x).$$

(2) If $a \in \mathbb{F}$ is a zero of $f(x)$ and $f^{(k)}(a) = 0$ $(0 \le k \le m-1)$ but $f^{(m)}(a) \neq 0$ for some $m \ge 1$, then show that the multiplicity of a in $f(x)$ is m, that is, show that

$$(x-a)^m | f(x) \text{ but } (x-a)^{m+1} \nmid f(x).$$

Corollary 5.13. *Let $\mathbb{F}$ be a finite field and $\mathbb{F}^\times = \mathbb{F} \setminus \{0\}$ be its multiplicative group. Then $\mathbb{F}^\times$ is a cyclic group.*

Proof. Since $\mathbb{F}^\times$ is a finite abelian group, the structure theorem for finitely generated abelian groups says that $\mathbb{F}^\times$ is isomorphic to

$$\mathbb{Z}_{m_1} \times \mathbb{Z}_{m_2} \times \cdots \times \mathbb{Z}_{m_r}$$

for some positive integers $m_1, m_2, \ldots, m_r$. An element $x \in \mathbb{F}^\times$ corresponds to $(a_1, a_2, \ldots, a_r)$ in $\mathbb{Z}_{m_1} \times \mathbb{Z}_{m_2} \times \cdots \times \mathbb{Z}_{m_r}$. Using the multiplication as our binary operation, $a_i \in \mathbb{Z}_{m_i}$ satisfies $a_i^{m_i} = 1$. We let $m = \operatorname{lcm}(m_1, m_2, \ldots, m_r)$, which is less than or equal to $m_1 m_2 \cdots m_r$. Then, for each $x \in \mathbb{F}^\times$, we know that the corresponding element of x^m satisfies that $(a_1, a_2, \ldots, a_r)^m = (a_1^m, a_2^m, \ldots, a_r^m) = (1, 1, \ldots, 1)$. That is, for any $x \in \mathbb{F}^\times$, $x^m = 1$ holds. Now we consider the polynomial

$$f(x) = x^m - 1 \in \mathbb{F}[x].$$

Note that the number of zeros of $f(x)$ in $\mathbb{F}$ is $|\mathbb{Z}_{m_1} \times \mathbb{Z}_{m_2} \times \cdots \times \mathbb{Z}_{m_r}| = m_1 m_2 \cdots m_r$, which is less than or equal to $m = \deg f(x)$ by the above Corollary. Therefore, we must have $m_1 m_2 \cdots m_r = m$, that is, $\gcd(m_i, m_j) = 1$ for each pair i, j, so that $\mathbb{F}^\times$ is isomorphic to $\mathbb{Z}_m$, a cyclic group. □

Remark 5.14. (1) A finite multiplicative group of an infinite field is also a cyclic group. For example, inside the field the complex numbers, the finite multiplicative group $U_n = \{z \in \mathbb{C} |\ z^n = 1\}$ of order n is isomorphic to $(\mathbb{Z}_n, +)$.

(2) $\mathbb{Z}_5^\times = \langle 2 \rangle$, $\mathbb{Z}_7^\times = \langle 3 \rangle$, $\mathbb{Z}_{11}^\times = \langle 2 \rangle$, $\mathbb{Z}_{13}^\times = \langle 2 \rangle$, $\mathbb{Z}_{17}^\times = \langle 3 \rangle$, $\mathbb{Z}_{19}^\times = \langle 2 \rangle$, $\mathbb{Z}_{23}^\times = \langle 5 \rangle$. However, an explicit generator of $\mathbb{Z}_p^\times$ (p, a prime), i.e., a primitive root of p, is not easy to find, when p is large. In many cases, $\mathbb{Z}_p^\times = \langle 2 \rangle$.[26]

[26] Refer to a table in [오이, Appendix]. For details on primitive roots, refer to [오이, Ch.5] or [NZM, §2.8].

(3) We'll see that there is a finite field $\mathbb{F}_{p^n}$ in Ch.6 §3. The multiplicative group $\mathbb{F}_{p^n}^\times$ is a cyclic group of order $p^n - 1$.

Now we want to determine whether a given polynomial is irreducible over $\mathbb{F}$ or not. This is a less hard problem rather than the factorization.

DEFINITION 5.15. Let $\mathbb{F}$ be a field. A polynomial $f(x) \in \mathbb{F}[x]$ of $\deg f(x) \geq 1$ is called *irreducible* over $\mathbb{F}$ if, for any decomposition $f(x) = g(x)h(x)$, we have $\deg g(x) = \deg f(x)$ or $\deg h(x) = \deg f(x)$.

EXAMPLE 5.16. A polynomial $f(x) \in \mathbb{F}[x]$ of $\deg f(x) = 2, 3$ is reducible over $\mathbb{F}$ if and only if $f(x)$ has a zero, that is, $f(x)$ has a linear factor over $\mathbb{F}$.

EXERCISE 5.17. (1) Show that for p a prime, the polynomial $x^p + a \in \mathbb{Z}_p[x]$ is not irreducible for any $a \in \mathbb{Z}_p$. [Hint: Fermat's little theorem]

(2) Find the number of irreducible polynomials of the form $x^2 + ax + b$ over $\mathbb{Z}_p$ (p, a prime).

(3) Find all primes p such that $x + 2$ is a factor of $x^4 + x^3 + x^2 - x + 1 \in \mathbb{Z}_p[x]$.

EXAMPLE 5.18. Over $\mathbb{F}_2 = \{0, 1\}$, the irreducible polynomials of degree 2 or 3 are $x^2 + x + 1$, $x^3 + x + 1$, $x^3 + x^2 + 1$.

EXERCISE 5.19. (1) Find all monic irreducible polynomials of degree 2 over $\mathbb{Z}_3$.

(2) Find all irreducible polynomials of degree 3 in $\mathbb{Z}_3[x]$.

EXERCISE 5.20. Determine whether each polynomial is irreducible. If it is reducible, write it as the product of irreducible polynomials.

(1) $3x^2 + x + 4 \in \mathbb{Z}_7[x]$
(2) $2x^2 + x + 3 \in \mathbb{Z}_5[x]$
(3) $x^3 + x^2 + x + 1 \in \mathbb{Z}_2[x]$
(4) $x^2 + x + 4 \in \mathbb{Z}_{11}[x]$
(5) $x^3 + 6 \in \mathbb{Z}_7[x]$
(6) $x^5 + 4x^4 + 4x^3 - x^2 - 4x + 1 \in \mathbb{Z}_5[x]$
(7) $x^4 + 4 \in \mathbb{Z}_5[x]$
(8) $x^4 + 1 \in \mathbb{Z}_p[x]$ (p, a prime)

Lemma 5.21 (Gauss). *Let $f(x) \in \mathbb{Z}[x]$. Then $f(x)$ factors into two polynomials of lower degrees r and s in $\mathbb{Q}[x]$ if and only if it is a product of two polynomials of degrees r and s in $\mathbb{Z}[x]$.*

Corollary 5.22. *Let* $f(x) = x^n + a_{n-1}x^{n-1} + \cdots + a_0 \in \mathbb{Z}[x]$*. If* $f(x)$ *has a zero in* $\mathbb{Q}$*, then it has a zero* $m \in \mathbb{Z}$*, which is a divisor of* a_0*.*

Proof. Since $f(x)$ has a linear factor in $\mathbb{Q}[x]$, Gauss' lemma implies that it has a linear factor $x - m$ in $\mathbb{Z}[x]$. So $f(x) = (x-m)(x^{n-1} + \cdots - a_0/m) \in \mathbb{Z}[x]$. □

EXAMPLE 5.23. A polynomial $x^2 - 2 \in \mathbb{Q}[x]$ is irreducible over $\mathbb{Q}$, since it has no zero in $\mathbb{Z}$ ($\pm 1, \pm 2$ are not zeros of $x^2 - 2$).

EXERCISE 5.24. (1) Demonstrate that $x^4 - 22x^2 + 1$ is irreducible over $\mathbb{Q}$.

(2) Let $f(x) = a_nx^n + a_{n-1}x^{n-1} + \cdots + a_1x + a_0 \in \mathbb{Z}[x]$ with $a_n \neq 0$. Without using Gauss' Lemma, show that if $f\left(\frac{r}{s}\right) = 0$, where r, s are relatively prime integers, then $r|a_0$ and $s|a_n$.

Solution. (2) From $f\left(\frac{r}{s}\right)s^{n-1} = 0$, we get that $\frac{a_nr^n}{s} \in \mathbb{Z}$, so s must be a divisor of a_n. Also, the condition that $f\left(\frac{r}{s}\right)\frac{s^n}{r} = 0$ induces that $\frac{a_0s^n}{r} \in \mathbb{Z}$, so r must be a divisor of a_0. □

Lemma 5.25 (Eisenstein's criterion[27]). *Let* $f(x) = a_nx^n + \cdots + a_0 \in \mathbb{Z}[x]$ *and* $p \in \mathbb{Z}$ *be a prime. If* $a_n \not\equiv 0 \pmod p$ *but* $a_i \equiv 0 \pmod p$ *for* $i < n$ *with* $a_0 \not\equiv 0 \pmod{p^2}$*, then* $f(x)$ *is irreducible over* $\mathbb{Q}$*.*

Proof. By Gauss' lemma, we show that $f(x)$ does not factor in $\mathbb{Z}[x]$. Suppose that

$$f(x) = (b_rx^r + \cdots + b_0)(c_sx^s + \cdots + c_0)$$

with $b_r, c_s \neq 0$ and $r, s < n$. The conditions say that $p \nmid b_rc_s$ and $p \mid b_0c_0$ but $p^2 \nmid b_0c_0$, so $p \nmid b_r$, $p \nmid c_s$, and $p \nmid b_0$, $p \mid c_0$. Let m be the smallest integer k such that $p \nmid c_k$, that is, $p \nmid c_m$, $p \mid c_{m-1}$, $\ldots$, $p \mid c_0$. Then $a_m = b_0c_m + b_1c_{m-1} + \cdots + b_mc_0$, so $p \nmid a_m$, thus we get $m = n$, which contradicts that $s < n$. □

EXAMPLE 5.26. (1) A polynomial $25x^5 - 9x^4 - 3x^2 - 12$ is irreducible over $\mathbb{Q}$ by taking $p = 3$ in Eisenstein's criterion.

(2) For any prime p, the polynomial

$$\Phi_p(x) = \frac{x^p - 1}{x - 1} = x^{p-1} + x^{p-2} + \cdots + x + 1$$

is irreducible over $\mathbb{Q}$, since $\Phi_p(x+1) = x^{p-1} + px^{p-2} + \binom{p}{2}x^{p-3} + \cdots + p$ is irreducible over $\mathbb{Q}$ by Eisenstein's criterion.

---

[27] Ferdinand Gotthold Max Eisenstein(1823–1852), a German mathematician.

EXERCISE 5.27. Determine whether each polynomial is irreducible over $\mathbb{Q}$.

(1) $x^4 + 1$

(2) $x^5 + 9x^4 + 12x^2 + 6$

(3) $x^4 + x + 1$

(4) $x^4 + 3x^2 + 3$

(5) $x^5 + 5x^2 + 1$

(6) $8x^3 - 6x + 1$

(7) $\frac{5}{2}x^5 + \frac{9}{2}x^4 + 15x^3 + \frac{3}{7}x^2 + 6x + \frac{3}{14}$

(8) $x^3 + 3x^2 - 8$

(9) $x^4 - 22x^2 + 1$

(10) $8x^3 + 6x^2 - 9x + 24$

REMARK 5.28. Note that $\mathbb{Z}$ is a unique factorization domain(UFD, for short), that is, an integral domain such that every element is a product of irreducible factors, uniquely up to ordering of irreducibles and units. In general, Gauss' lemma and Eisenstein's criterion hold for any UFD and its quotient field.

The following theorem says that any irreducible polynomial in $\mathbb{F}[x]$ is a prime element, so $\mathbb{F}[x]$ is a unique factorization domain. The proof will be given in Ch.5 Theorem 4.9.

Theorem 5.29. *Let $\mathbb{F}$ be a field.*

(a) *If any irreducible $p(x) \in \mathbb{F}[x]$ divides $r(x)s(x)$ for $r(x), s(x) \in \mathbb{F}[x]$, then $p(x)|r(x)$ or $p(x)|s(x)$.*

(b) *Any polynomial $f(x) \in \mathbb{F}[x]$ of positive degree factors into a product of irreducible polynomials, uniquely up to ordering of irreducibles and units.*

In Chapter 5 §6, we will see that $\mathbb{F}[x]$ is a principal ideal domain, i.e., an integral domain where each ideal is generated by one element. In fact, we can find the greatest common divisor among the polynomials, using the Euclidean algorithm.

REMARK 5.30. * In the ring $\mathbb{Q}[x, y]$ of polynomials with two variables over $\mathbb{Q}$, consider a polynomial $xy^2 - x$. Divide it by $xy - y$ and divide the remainder by $y^2 - x$. Finally, we get 0 as the remainder:

$$xy^2 - x \overset{xy-y}{\longrightarrow} y^2 - x \overset{y^2-x}{\longrightarrow} 0.$$

However, changing the order of division, we obtain that

$$xy^2 - x \xrightarrow{y^2-x} x^2 - x \xrightarrow{xy-y} x^2 - x.$$

The results are different: the multivariate division algorithm is dependent on the order of each division. In Chapter 8, we will solve this "bad" phenomenon.

Midterm Exam

(April 24, 2014)

1. (30 points) Determine whether each statement is true or not: True(T) or False(F). (If you answer correctly, you get 3 points; If nothing, 1 point; If wrong, 0 point.)
   1) The set $M_{n\times n}(\mathbb{Q})$ of all $n \times n$ matrices with entries in $\mathbb{Q}$ is a ring under the matrix addition and multiplication.
   2) There are infinitely many ring homomorphisms $\phi : \mathbb{Z} \to \mathbb{Z}$.
   3) $\mathbb{C}$ is isomorphic to $\left\{ \begin{pmatrix} a & b \\ -b & a \end{pmatrix} \mid a, b \in \mathbb{R} \right\}$ as a ring.
   4) Every division ring is commutative.
   5) The ring $\mathbb{Z}_p$ (p, a prime) has no zero divisors.
   6) Every integral domain is a field.
   7) The characteristic of an integral domain must be either 0 or a prime.
   8) $\mathbb{Q}$ is the quotient field of $\mathbb{Z}$.
   9) For a finite field $\mathbb{F}$, $\mathbb{F}^\times = \mathbb{F} \setminus \{0\}$ with multiplication is a cyclic group.
   10) For any prime p, the polynomial $x^{p-1} + x^{p-2} + \cdots + x + 1$ is irreducible over $\mathbb{Q}$.

2. (20pts) (1) Let S be a set and $P(S)$ be the power set of S, i.e., the set of all subsets of S. Define two binary operations $+$ and $\times$ for $A, B \in P(S)$ so that $P(S)$ is a Boolean ring.
   (2) Give the definition of an integral domain.
   (3) Give an example of a commutative ring with 1 which is not an integral domain.
   (4) State Wilson's theorem.
3. (20pts) (1) Over $\mathbb{F}_2 = \{0, 1\}$, find all irreducible polynomials of degree 3.
   (2) State Eisenstein's criterion for the irreducibility of $a_n x^n + a_{n-1}x^{n-1} + \cdots + a_1 x + a_0 \in \mathbb{Z}[x]$.

- For each of the following problems, you should justify your answer.

4. (20pts) (1) Find all solutions of $45x \equiv 15 \pmod{24}$.
   (2) Let $f_1(x) = x^4 - 2x^3 - x^2 + 2x$ and $f_2(x) = x^3 - 7x^2 + 14x - 8$. Find $\gcd(f_1, f_2)$.
5. (10pts) Let $\mathbb{F}_p$ be a finite field with p elements (p, a prime). Then show that

$$x^{p-1} - 1 = (x-1)(x-2)\cdots(x-p+1) \in \mathbb{F}_p[x].$$

Midterm Exam

(April 26, 2012)

1. (10 points) Write Fermat's little theorem, and Wilson's theorem.
2. (10pts) What does it mean that two positive real numbers are commensurable?
3. (10pts) Compute the quotient and remainder when $3x^4 + x^3 + 2x^2 + 1$ is divided by $x^2 + 4x + 2$ in $\mathbb{Z}_5[x]$.

- For the following 3 problems, you should justify your answer.

4. (10pts) Determine whether the ring $(M_{2\times 2}(\mathbb{Q}), +, \cdot\,)$ is an integral domain or not.
5. (20pts) (1) Solve the linear congruence $15x \equiv 9 \pmod{18}$.
   (2) Solve the congruences $x^2 \equiv -1 \pmod{13}$ and $x^2 \equiv -1 \pmod{47}$.
6. (10pts) Find a generator of the cyclic group $\mathbb{Z}_{17}^{\times}$.

7. (30pts) Determine whether each statement is true or not: True(T) or False(F).
   (If you answer correctly, you get 3 points; If nothing, 1 point; If wrong, 0 point.)
   1) The rings $(\mathbb{Z}, +, \cdot\,)$ and $(2\mathbb{Z}, +, \cdot\,)$ are isomorphic.
   2) The set of quaternions $\mathbb{H} = \{a + bi + cj + dk \mid a, b, c, d \in \mathbb{R}\}$ with $i^2 = j^2 = k^2 = ijk = -1$ is a commutative ring.
   3) $(\mathbb{Z}_{pq}, +, \cdot)$ $(p, q$, primes) is an integral domain.
   4) If D is an integral domain then D is a field.
   5) Let S be a set with $|S| \geq 2$. Then the Boolean ring $P(S)$ with the symmetric difference $+$ and the intersection $\cdot$ is an integral domain.
   6) The characteristic of an integral domain cannot be a prime.
   7) The quotient field of an integral domain D is the smallest field containing D.
   8) The ring $\mathbb{Q}[x]$ of polynomials with coefficients in $\mathbb{Q}$ is an integral domain.
   9) For each prime p, the polynomial $x^{p-1} + x^{p-2} + \cdots + x + 1$ is irreducible over $\mathbb{Q}$.
   10) I understand the materials in the class over 70%.

Midterm Exam

(April 22, 2011)

1. Determine whether each statement is true or not: True(T) or False(F).
   (If you answer correctly, you get 3 points; If nothing, 1 point; If wrong, 0 point.)
   1) There is a root formula for quintic equations.
   2) Two rings $2\mathbb{Z}$ and $3\mathbb{Z}$ are isomorphic.
   3) The ring $\mathbb{Z}[\sqrt{2}] = \{a + b\sqrt{2} |\ a, b \in \mathbb{Z}\}$ is a field.
   4) Any finite integral domain is a field.
   5) The characteristic of an integral domain must be a prime.
   6) For a prime p and a positive integer which is not a multiple of p, we have $a^{p-1} \equiv 1 \pmod{p}$.
   7) For $n = p - 1$ with a prime p, no integer of the form $n! + 1$ is a prime.
   8) Any polynomial $f(x) \in \mathbb{F}[x]$ ($\mathbb{F}$, a field) of degree n has at most n zeros.
   9) For a finite field $\mathbb{F}$, the multiplicative group $\mathbb{F}^\times$ is cyclic.
   10) Any p-th cyclotomic polynomial $\Phi_p(x) = x^{p-1} + x^{p-2} + \cdots + x + 1$ (p, a prime) is irreducible over $\mathbb{Q}$.

- For each of the following problems, you should justify your answer.

2. (15 points) Let $\mathbb{Z}[\sqrt{2}] = \{a + b\sqrt{2} |\ a, b \in \mathbb{Z}\}$ and $H = \left\{ \begin{pmatrix} a & 2b \\ b & a \end{pmatrix} |\ a, b \in \mathbb{Z} \right\}$. Show that $\mathbb{Z}[\sqrt{2}]$ is isomorphic to H as rings.
3. (20pts) (1) On the set $P(S)$ of a set S with $|S| \geq 2$, give a Boolean ring structure, that is, define the addition $+$ and multiplication $\cdot$ on $P(S)$ such that $P(S)$ is a Boolean ring.
   (2) What is the additive identity of the Boolean ring $P(S)$?
   (3) Find the characteristic of the Boolean ring $P(S)$. Is $P(S)$ an integral domain?
4. (15pts) (1) Define $R^\times$ to be the set of all units in a ring R. Find $\mathbb{Z}_{20}^\times$.
   (2) Solve a linear congruence $15x \equiv 10 \pmod{20}$.
5. (20pts) (1) Find all monic irreducible polynomials of degree 2 over $\mathbb{Z}_3$.
   (2) Determine whether $x^5+9x^4+12x^2+6$ is irreducible over $\mathbb{Q}$. [Hint: Eisenstein's criterion.]

6. The quaternions *

Consider the set

$$\mathbb{H} = \{a + bi + cj + dk \mid a, b, c, d \in \mathbb{R}\} = \{\alpha + \beta j \mid \alpha, \beta \in \mathbb{C}\}$$

of quaternions, equipped with the usual addition and the multiplication by

$$i^2 = j^2 = k^2 = ijk = -1.$$

We call it the *Hamiltonian skew-field.*[28] This is useful when we describe a rotation in the real 3-dimensional space.[29] Note that the multiplication in $\mathbb{H}$ is not commutative.

EXERCISE 6.1. Find the multiplication table among i, j, k.

Our question is:

How to describe a rotation in $\mathbb{R}^3$ using the quaternions?

Let $\mathbb{H} = \mathbb{R} \oplus V$ where

$$V = \{xi + yj + zk \mid x, y, z \in \mathbb{R}\}$$

is the set of quaternions with real part 0. Note that V is isomorphic to $\mathbb{R}^3$ as a vector space over $\mathbb{R}$.

For 3-dimensional vectors $\mathbf{v_1}, \mathbf{v_2} \in V \cong \mathbb{R}^3$, Gibbs[30] defined

$$\mathbf{v_1}\mathbf{v_2} := -\mathbf{v_1} \cdot \mathbf{v_2} + \mathbf{v_1} \times \mathbf{v_2} \in \mathbb{R} \oplus V.$$

In particular, if $\mathbf{v_1} \perp \mathbf{v_2}$ then $\mathbf{v_1}\mathbf{v_2} = -\mathbf{v_2}\mathbf{v_1}$.

PROPERTY 6.2. For $q = t + \mathbf{v} = t + xi + yj + zk \in \mathbb{H}$, the conjugate of q is defined to be

$$\overline{q} = t - \mathbf{v} = t - xi - yj - zk,$$

and the absolute value of q is

$$|q| = \sqrt{t^2 + |\mathbf{v}|^2} = \sqrt{t^2 + x^2 + y^2 + z^2}.$$

---

[28]This was discovered by William Rowan Hamilton(1805–1865), an Irish physicist and mathematician. On 16 October 1843, he was out walking along the Royal Canal in Dublin with his wife when the solution in the form of the equation $i^2 = j^2 = k^2 = ijk = -1$ suddenly occurred to him, promptly carving this equation using his penknife into the side of the nearby Brougham Bridge.

[29]For details, see `http://www.math.snu.ac.kr/~hongjong/잡필/QUATERNION.pdf`. Also, it is interesting to read `http://navercast.naver.com/science/math/4132, -/4788, -/5036`.

[30]Josiah Willard Gibbs(1839–1903), a theoretical physicist. He is the 1st American engineering doctor.

Let $q_1, q_2, q_3, q \in \mathbb{H}$. Then the following hold:

1) $(q_1 + q_2)q_3 = q_1q_3 + q_2q_3, \quad q_1(q_2 + q_3) = q_1q_2 + q_1q_3, \quad (q_1q_2)q_3 = q_1(q_2q_3)$.
2) $q\overline{q} = |q|^2 = \overline{q}q$, so $q^{-1} = \frac{\overline{q}}{|q|^2}$ if $q \neq 0$.
3) $\overline{q_1q_2} = \overline{q_2}\,\overline{q_1}, \quad |q_1q_2| = |q_1||q_2|$.

EXERCISE 6.3. Show the above properties.

For each $q \in \mathbb{H}$, we define a function on V

$$M_q : V \to V, \ w \mapsto qw\overline{q}.$$

We have to check that $qw\overline{q}$ does belong to $V \subset \mathbb{H}$. This is OK, since

$$\overline{qw\overline{q}} = \overline{\overline{q}}\ \overline{w}\ \overline{q} = q(-w)\overline{q} = -qw\overline{q}.$$

Proposition 6.4. *The function M_q is a linear map which is orientation-preserving, and is also length-preserving if $|q| = 1$.*

Proof. We can easily check that, for each $w_1, w_2, w \in V$ and $k \in \mathbb{R}$,

$$M_q(w_1 + w_2) = M_q(w_1) + M_q(w_2), \quad M_q(kw) = kM_q(w).$$

This linear map M_q is written as a matrix. It has a positive determinant, which is left for the reader.

We check that $|M_q(w)| = |q|^2|w|$. □

EXAMPLE 6.5. We identify V with $\mathbb{R}^3$ via $xi + yj + zk \leftrightarrow (x, y, z)$, and denote by $[M_q]$ the matrix of M_q with respect to the standard basis $\mathcal{E} = \{i, j, k\}$ of V. In particular,

$$[M_i] = \begin{pmatrix} 1 & 0 & 0 \\ 0 & -1 & 0 \\ 0 & 0 & -1 \end{pmatrix}, \quad [M_j] = \begin{pmatrix} -1 & 0 & 0 \\ 0 & 1 & 0 \\ 0 & 0 & -1 \end{pmatrix}, \quad [M_k] = \begin{pmatrix} -1 & 0 & 0 \\ 0 & -1 & 0 \\ 0 & 0 & 1 \end{pmatrix}.$$

Note that M_i is the rotation by π along the i-axis in $V \cong \mathbb{R}^3$.

EXERCISE 6.6. For any $q = t + xi + yj + zk \in \mathbb{H}$, what is the 3×3 matrix $[M_q]$ representing M_q?

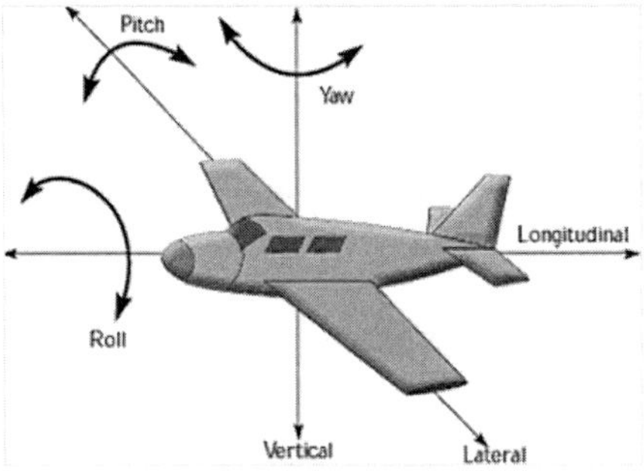

Denote the set of unit quaternions by

$$S_3 := \{q \in \mathbb{H} |\ |q| = 1\},$$

and define

$$SO_3 := \{M_q |\ q \in S_3\}.$$

Consider a mapping

$$M : S_3 \to SO_3 \ \text{ by } \ M(q) = M_q.$$

Note that M is 2-1, since $M_q = M_{-q}$.

REMARK 6.7. The set S_3 in $\mathbb{H}$ is closed under the multiplication. Note that S_3 is a multiplicative group and so is $S_3/\{\pm 1\}$. Also, SO_3 is closed under the composition, since

$$M_{q_2} \circ M_{q_1} = M_{q_2 q_1}.$$

We see that $S_3/\{\pm 1\} \cong SO_3$ as a group.

In order to see what is the meaning of M_q. we write $q \in S_3$ as

$$q = t + \mathbf{v} = \cos\theta + \mathbf{u}\sin\theta,$$

where $\mathbf{v}/|\mathbf{v}| = \mathbf{u} \in V$ (let $\mathbf{u} = \mathbf{0}$ if $\mathbf{v} = \mathbf{0}$), $-1 \le t = \cos\theta \le 1$, $0 \le |\mathbf{v}| = \sin\theta \le 1$ with $0 \le \theta \le \pi$.

If $\mathbf{v} = \mathbf{0}$, that is, $q = \pm 1$, then $M_1 = M_{-1}$ is the identity map on $V \cong \mathbb{R}^3$.

Assume that $q \ne \pm 1$ and

$$\tfrac{\mathbf{v}}{|\mathbf{v}|} = \mathbf{u} \in V, \ 0 \le t = \cos\theta < 1, \ 0 < |\mathbf{v}| = \sin\theta \le 1 \text{ with } 0 < \theta \le \tfrac{\pi}{2}.$$

EXAMPLE 6.8. (1) $\frac{1}{\sqrt{2}} + \frac{1}{\sqrt{2}}i = \cos\frac{\pi}{4} + i\sin\frac{\pi}{4}$.

(2) $\frac{1}{\sqrt{2}} + \frac{1}{2}i + \frac{1}{2}j = \cos\frac{\pi}{4} + \left(\frac{1}{\sqrt{2}}i + \frac{1}{\sqrt{2}}j\right)\sin\frac{\pi}{4}$.

(3) $\frac{1}{2} + \frac{1}{2}i + \frac{1}{2}j + \frac{1}{2}k = \cos\frac{\pi}{3} + \left(\frac{1}{\sqrt{3}}i + \frac{1}{\sqrt{3}}j + \frac{1}{\sqrt{3}}k\right)\sin\frac{\pi}{3}$.

Theorem 6.9. *Let $q = \cos\theta + \mathbf{u}\sin\theta \in S_3 \subset \mathbb{H}$ as above. Then M_q is the rotation by 2θ along the $\mathbf{u}$-axis in $V \cong \mathbb{R}^3$.*

Proof. We calculate that

$$M_q(\mathbf{u}) = (\cos\theta + \mathbf{u}\sin\theta)\mathbf{u}(\cos\theta - \mathbf{u}\sin\theta) = \mathbf{u}.$$

By Proposition 6.4, M_q is the rotation along the $\mathbf{u}$-axis.

To find the rotation angle, we let $\mathbf{w}$ to be a unit vector orthogonal to $\mathbf{u}$ and consider the plane spanned by $\mathbf{w}$ and $\mathbf{uw} = \mathbf{u} \times \mathbf{w}$. We can check that

$$M_q(\mathbf{w}) = (\cos\theta + \mathbf{u}\sin\theta)\mathbf{w}(\cos\theta - \mathbf{u}\sin\theta) = \mathbf{w}\cos 2\theta + \mathbf{uw}\sin 2\theta,$$

and

$$M_q(\mathbf{uw}) = -\mathbf{w}\sin 2\theta + \mathbf{uw}\cos 2\theta = \mathbf{w}\cos\left(\tfrac{\pi}{2} + 2\theta\right) + \mathbf{uw}\sin\left(\tfrac{\pi}{2} + 2\theta\right).$$

Hence we proved. □

Corollary 6.10. *The composition of two rotations in $V \cong \mathbb{R}^3$ is also a rotation in V. We know*

$$M_{q_2} \circ M_{q_1} = M_{q_2 q_1},$$

that is, $q_2 q_1 \in S_3$ indicates the rotation axis and the angle. More precisely,

$$\begin{aligned}
&(\cos\theta_2 + \mathbf{u_2}\sin\theta_2)(\cos\theta_1 + \mathbf{u_1}\sin\theta_1) \\
&= \cos\theta_2\cos\theta_1 - \mathbf{u_2}\cdot\mathbf{u_1}\sin\theta_2\sin\theta_1 \\
&\quad + \mathbf{u_2}\sin\theta_2\cos\theta_1 + \mathbf{u_1}\sin\theta_1\cos\theta_2 + \mathbf{u_2}\times\mathbf{u_1}\sin\theta_2\sin\theta_1 \\
&= \cos\alpha + \mathbf{u}\sin\alpha,
\end{aligned}$$

for a unique $\mathbf{u} \in V_1 = \{\mathbf{v} \in V \mid |\mathbf{v}| = 1\}$ and $0 \le \alpha \le \pi$.

Remark 6.11. (1) We can easily find the inverse transformation of M_q, that is,

$$M_q^{-1} = M_{\overline{q}}$$

for $q \in S_3$.

(2) We define the special orthogonal group of degree 3 to be

$$SO(3) = \{A \in M_{3\times 3}(\mathbb{R}) \mid A^T A = I,\ \det A = 1\},$$

which is just called the group of rotations in $\mathbb{R}^3$. Note that $SO(3)$ is isomorphic to SO_3 as a group, and that

$$M_{q_2 q_1} \neq M_{q_1 q_2} \quad \text{for } q_1, q_2 \in S_3.$$

(3) Given $\mathbf{v}$ and $\mathbf{w}$ in $V = \{xi + yj + zk|\ x, y, z \in \mathbb{R}\}$ with $|\mathbf{v}| = |\mathbf{w}|$ and $\mathbf{v} \neq \pm\mathbf{w}$, what is a rotation transforming $\mathbf{v}$ into $\mathbf{w}$?

Let $|\mathbf{v}| = |\mathbf{w}| = 1$ for convenience, and let $0 < 2\theta < \pi$ be the angle between $\mathbf{v}$ and $\mathbf{w}$. Then

$$\mathbf{u} := \frac{\mathbf{v} \times \mathbf{w}}{|\mathbf{v} \times \mathbf{w}|} = \frac{\mathbf{v} \times \mathbf{w}}{\sin 2\theta}$$

is a unit vector orthogonal to $\mathbf{v}$ and $\mathbf{w}$. For $q = \cos\theta + \mathbf{u}\sin\theta$, the transformation $M_q : V \to V$ is a rotation by 2θ along the $\mathbf{u}$-axis, which satisfies $M_q(\mathbf{v}) = \mathbf{w}$.

The following exercise reveals a connection between the quaternions and the complex numbers.

EXERCISE 6.12. Define $\phi : \mathbb{H} \to M_{2\times 2}(\mathbb{C})$ by

$$\phi(\alpha + \beta j) = \begin{pmatrix} \alpha & \beta \\ -\overline{\beta} & \overline{\alpha} \end{pmatrix}.$$

(1) Show that, for each $q_1, q_2 \in \mathbb{H}$, it holds that

$$\phi(q_1 + q_2) = \phi(q_1) + \phi(q_2), \quad \phi(q_1 q_2) = \phi(q_1)\phi(q_2).$$

[Hint: Be cautious that $\beta j = j\overline{\beta}$.]

(2) Show that ϕ is injective, i.e., 1-1, so

$$\mathbb{H} \cong \left\{ \begin{pmatrix} \alpha & \beta \\ -\overline{\beta} & \overline{\alpha} \end{pmatrix} \in M_{2\times 2}(\mathbb{C})|\ \alpha, \beta \in \mathbb{C} \right\}$$

under the addition and multiplication. Thus, a beautiful description for the quaternions using the 2×2 complex matrices is

$$a\begin{pmatrix} 1 & 0 \\ 0 & 1 \end{pmatrix} + b\begin{pmatrix} i & 0 \\ 0 & -i \end{pmatrix} + c\begin{pmatrix} 0 & 1 \\ -1 & 0 \end{pmatrix} + d\begin{pmatrix} 0 & i \\ i & 0 \end{pmatrix}$$

with $a, b, c, d \in \mathbb{R}$.

REMARK 6.13. * Quaternions are used to describe Maxwell's electromagnetic equations and they are also useful in quantum mechanics.[31]

Now we have a question. Is there another meaningful number system?

An algebra is a vector space equipped with the multiplication. We introduce the 8-dimensional algebra over $\mathbb{R}$ with basis

$$\{1, e_1, e_2, e_3, e_4, e_5, e_6, e_7\},$$

[31] For applications in Physics, refer to a material, for instance, the website by Doug Sweetser; `http://world.std.com/~sweetser/quaternions/qindex/qindex.html`.

where the multiplication is as follows:

- $e_1, \ldots, e_7$ are square roots of -1;
- e_i and e_j are anticommutative when $i \neq j$, that is, $e_i e_j = -e_j e_i$;
- the index cycling identity holds, that is, $e_i e_j = e_k$ implies $e_{i+1} e_{j+1} = e_{k+1}$ where the index is in $\mathbb{Z}_7$;
- the index doubling identity holds, that is, $e_i e_j = e_k$ implies $e_{2i} e_{2j} = e_{2k}$;
- we set $e_1 e_2 = e_4$.

This 8-dimensional $\mathbb{R}$-algebra is called the *octonion* algebra or the Cayley algebra[32], denoted by $\mathbb{O}$. Note that the multiplication in $\mathbb{O}$ is not associative. As a picture, we depict the multiplication on the Fano[33] plane.

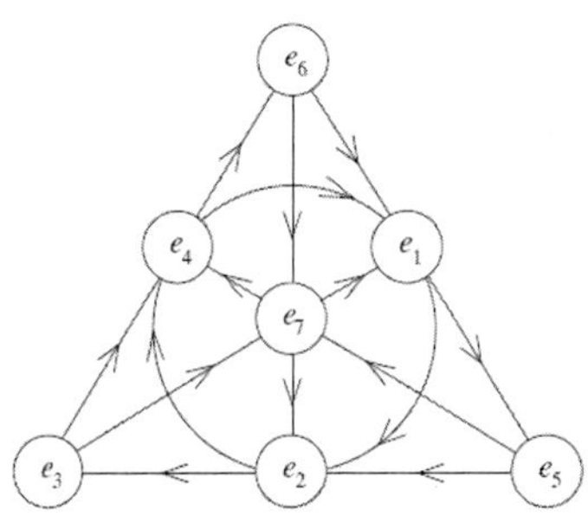

FIGURE 1. The Fano plane in $\mathbb{Z}_2^3$

Explicitly,

$$\mathbb{O} = \{A + lB |\ A, B \in \mathbb{H}\},$$

where $i = e_1$, $j = e_2$, $ij = k = e_4$, $l = e_7$, $li = e_3$, $lj = e_6$, $lk = e_5$.

REMARK 6.14. The conjugation $*$ on an algebra A is a map on A such that

$$a^{**} = a \quad \text{and} \quad (ab)^* = b^* a^*$$

for each $a, b \in A$. The conjugate of a complex number and a quaternion a is usually denoted by $\overline{a}$. We call an algebra with a conjugation $*$ to be a $*$-algebra.

[32] The algebra of octonions was discovered by Hamilton's college friend John T. Graves in December 1843. It was discovered independently by Arthur Cayley in 1845.

[33] Gino Fano(1871–1952), an Italian mathematician.

In general, we construct a new $*$-algebra $A' = A^2$ from a given $*$-algebra A. This process is called the *Cayley-Dickson*[34] *doubling construction*[35]. We have in mind how to construct

$$\mathbb{R} < \mathbb{C} < \mathbb{H} < \mathbb{O}.$$

Let A be a $*$-algebra and take any two elements (a,b), (c,d) in A^2. Define the multiplication on A^2 by

$$(a,b)(c,d) := (ac - db^*, a^*d + cb),$$

and the conjugation is defined to be

$$(a,b)^* := (a^*, -b).$$

Then A^2 is also a $*$-algebra.

DEFINITION 6.15. (a) An algebra A called a *division algebra* if every nonzero element $a \in A$ has a multiplicative inverse.

(b) The multiplication on A is called alternative if $(a^2)b = a(ab)$ for each $a, b \in A$.

(c) A $*$-algebra A is called nicely normed if $a + a^* \in \mathbb{R}$ and $aa^* = a^*a > 0$ for each nonzero $a \in A$.

We mention the following propositions and theorems without proof.

Proposition 6.16. $*$ *Let A' be a new $*$-algebra $A' = A^2$ from a $*$-algebra A via the Cayley-Dickson doubling construction.*

(a) *A' is never real.*

(b) *A is real (thus, A is commutative) if and only if A' is commutative.*

(c) *A is commutative and associative if and only if A' is associative.*

(d) *A is associative and nicely normed if and only if A' is alternative and nicely normed.*

(e) *A is nicely normed if and only if A' is nicely normed.*

Theorem 6.17. $*$

(a) (Frobenius[36] 1877) *The associative division algebras over $\mathbb{R}$ are only $\mathbb{R}$, $\mathbb{C}$ and $\mathbb{H}$.*

---

[34] Leonard Eugene Dickson(1874–1954), an American mathematician.

[35] See, for instance, the paper [Bz].

[36] Ferdinand Georg Frobenius(1849–1917), a German mathematician.

(b) (Hurwitz[37] 1898) *The normed division algebras over* $\mathbb{R}$ *are only* $\mathbb{R}$, $\mathbb{C}$, $\mathbb{H}$ *and* $\mathbb{O}$.

(c) (Adams[38] 1963) $\mathbb{R}^n$ *is a normed division algebra.*

$\iff n = 1, 2, 4, 8.$

$\iff$ *The maximal number of linearly independent vector fields on* S^{n-1} *is* $n-1$.

[37]Adolf Hurwitz(1859–1919), a German mathematician.

[38]John Frank Adams(1930–1989), a British mathematician. He is a founder of homotopy theory.

CHAPTER 5

Quotient rings

The aim of this chapter is to study the structure of a ring and how to construct a new ring by factoring an old one. In Ch.2, we learned the notion of a normal subgroup of a group and that of a quotient group. Its ring-theoretic version will be investigated.

1. Ideals

We begin with stating basic facts on ring homomorphisms. Recall that $\phi : R \to R'$ is a ring homomorphism if, for each $a, b \in R$,

$$\phi(a+b) = \phi(a) + \phi(b), \quad \text{and} \quad \phi(ab) = \phi(a)\phi(b).$$

PROPERTY 1.1. Let $\phi : R \to R'$ be a ring homomorphism.

(a) $\phi(0) = 0'$, and $\phi(-a) = -\phi(a)$ for each $a \in R$.

(b) If $S \leq R$ is a subring then $\phi(S) \leq R'$.

(c) If $S' \leq R'$ is a subring then $\phi^{-1}(S') \leq R$.

(d) Assume that $1 \in R$ and $\phi(1) \neq 0'$. Then $\phi(1) = 1' \in \phi(R)$.

EXERCISE 1.2. Check the above properties.

DEFINITION 1.3. Let $\phi : R \to R'$ be a ring homomorphism. The subset

$$\phi^{-1}(0') = \{r \in R | \ \phi(r) = 0'\}$$

is a subring of R, which is called the *kernel* of ϕ, denoted by $\ker \phi$.

Proposition 1.4. *Let $\phi : R \to R'$ be a ring homomorphism and $a \in R$. Then*

$$\phi^{-1}(\phi(a)) = a + \ker\phi = \ker\phi + a.$$

Proof. We can easily check that

$$x \in \phi^{-1}(\phi(a)) \Leftrightarrow \phi(x) = \phi(a) \Leftrightarrow (-\phi(a)) + \phi(x) = 0' \Leftrightarrow \phi((-a)+x) = 0'$$

$$\Leftrightarrow (-a) + x + \in \ker\phi \Leftrightarrow x \in a + \ker\phi.$$

Similarly, $\phi^{-1}(\phi(a)) = \ker\phi + a$ can be shown. □

Corollary 1.5. *A ring homomorphism $\phi : R \to R'$ is 1-1 if and only if $\ker\phi = \{0\}$.*

REMARK 1.6. A homomorphism $\phi : R \to R'$ is an isomorphism if it is 1-1 and onto. We say that R is isomorphic to R', denoted by $R \cong R'$. Note that $\cong$ is an equivalence relation.

Theorem 1.7. *Let $\phi : R \to R'$ be a ring homomorphism with $I = \ker\phi$. Then*

(a) *The set $R/I = \{a + I|\ a \in R\}$ is a ring with binary operations*

$$(a + I) + (b + I) := (a + b) + I, \quad (a + I)(b + I) := (ab) + I.$$

(b) (The fundamental homomorphism theorem) *Let $\gamma : R \to R/I$ be the canonical projection. If we define $\mu : R/I \to \phi(R)$ by $a + I \mapsto \phi(a)$, then μ is an isomorphism.*

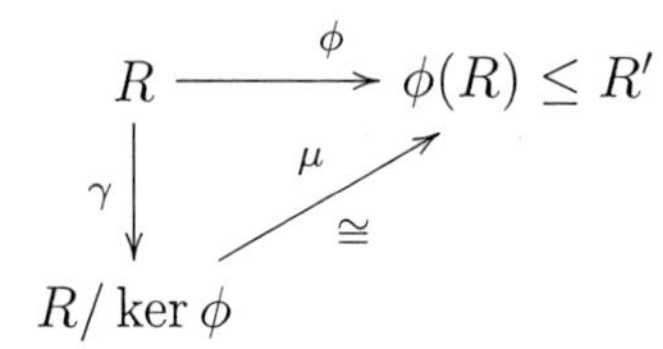

Proof. We have to check that the multiplication on R/I is well-defined. Let $a + k \in a + I$ ($k \in I = \ker\phi$) and $b + k' \in b + I$ ($k' \in I = \ker\phi$) be another elements in $a + I$ and $b + I$, that is, $a + k + I = a + I$ and $b + k' + I = b + I$. Then the multiplication is

$$(a + k)(b + k') + I = ab + ak' + kb + kk' + I = ab + I,$$

since $\phi(ak' + kb + kk') = \phi(a)\phi(k') + \phi(k)\phi(b) + \phi(kk') = 0$ implies $ak' + kb + kk' \in I$. So we proved. □

EXAMPLE 1.8. Consider the map $\phi : \mathbb{Z} \to \mathbb{Z}_n$ sending an integer to its remainder. Then $\ker\phi = n\mathbb{Z}$, and $\mathbb{Z}/n\mathbb{Z} \cong \mathbb{Z}_n$. For example, in $\mathbb{Z}/4\mathbb{Z}$,

$$(2 + 4\mathbb{Z})(3 + 4\mathbb{Z}) = 6 + 4\mathbb{Z} = 2 + 4\mathbb{Z}.$$

Let $I \leq R$ be a subring. State and prove the necessary and sufficient condition of I for R/I to be a quotient ring.

Theorem 1.9. *For a subring $I \leq R$, we define a multiplication on $R/I = \{a + I|a \in R\}$ to be*

$$(a + I)(b + I) = (ab) + I.$$

Then this operation is well-defined if and only if

$$ak \in I \quad \text{and} \quad kb \in I \quad \text{for each } a, b \in R \text{ and } k \in I.$$

Proof. $\Rightarrow$) Take any $a + k \in a + I$ and $b + k' \in b + I$ with $k, k' \in I$. Then we apply the well-definedness of the operation to the case of $k = 0$ and the case of $k' = 0$, leading us to the desired result.

$\Leftarrow$) For any $a, b \in R$ and any $k \in I$, since $ak' \in I$ and $kb \in I$, we get that

$$(a + k)(b + k') + I = ab + ak' + kb + kk' + I = ab + I.$$

So the multiplication is well-defined. □

DEFINITION 1.10. Let R be a ring.

(a) A subring I is called an *ideal* in R if $aI \subset I$ and $Ia \subset I$ for each $a \in R$.

(b) For an ideal I of R, a ring $R/I = \{a + I |\ a \in R\}$ with binary operations

$$(a + I) + (b + I) = (a + b) + I, \quad (a + I)(b + I) = (ab) + I$$

is called the *quotient ring* (or *factor ring*) *of R modulo I*.

EXAMPLE 1.11. (1) Any subring $n\mathbb{Z}$ is an ideal of $\mathbb{Z}$. Is $\mathbb{Z}$ an ideal of $\mathbb{Q}$?

(2) Let $R = M_{2\times 2}(\mathbb{Z})$ and $I = M_{2\times 2}(2\mathbb{Z})$. Then I is an ideal of R, and

$$R/I = \left\{ \begin{pmatrix} a & b \\ c & d \end{pmatrix} + I |\ a, b, c, d = 0, 1 \right\}.$$

Theorem 1.12. (a) *Let I be an ideal of R. Then the map $\gamma : R \to R/I$ defined by $x \mapsto x + I$ is a ring homomorphism with* $\ker \gamma = I$. *We call γ the* canonical projection *of R onto R/I.*

(b) *Let $\phi : R \to R'$ be a ring homomorphism with $I = \ker \phi$. Then the map $\mu : R/I \to \phi(R)$ defined by $a + I \mapsto \phi(a)$ is an isomorphism with $\phi = \mu \circ \gamma$, where $\gamma : R \to R/I$ is the canonical projection.*

To summarize up to now:

Let I be a subring of R. Then the following are equivalent:

(i) I is the kernel of a homomorphism ϕ from R.

(ii) The operation on $R/I = \{r + I |\ r \in R\}$ defined by $(a+I)(b+I) = (ab)+I$ is well-defined.

(iii) I is an ideal of R, that is, $RI \subseteq I$ and $IR \subseteq I$.

DEFINITION 1.13. Let R be a commutative ring with 1. For $a \in R$, we let

$$\langle a \rangle = \{ra |\ r \in R\},$$

which is an ideal of R, called the *principal ideal generated by* a. We say that the element a is a generator of the ideal $\langle a \rangle$ of R.

EXAMPLE 1.14. (1) Find $\langle 3 \rangle \subset \mathbb{Z}_{10}$, $\langle 4 \rangle \subset \mathbb{Z}_{10}$, $\langle 2 \rangle \subset \mathbb{Q}$, and $\langle x \rangle \subset \mathbb{R}[x]$, explicitly.

(2) Using the fundamental homomorphism theorem, we show that $\mathbb{R}[x]/\langle x \rangle \cong \mathbb{R}$.

(3) $\mathbb{R}[x]/\langle x-1 \rangle \cong \mathbb{R}$, $\mathbb{Q}[x]/\langle x^2-2 \rangle \cong \mathbb{Q}[\sqrt{2}]$. What is an explicit isomorphism for each of the preceding examples?

(4)(Cauchy 1847) We have an important isomorphism

$$\mathbb{R}[x]/\langle x^2+1 \rangle \cong \mathbb{C}.$$

(5) More generally, for $a_1, \ldots, a_n \in R$, we consider the (left) ideal generated by $a_1, \ldots, a_n$,

$$\langle a_1, \ldots, a_n \rangle = \{\sum_{i=1}^{n} r_i a_i |\ r_i \in R\}.$$

What is $\langle x, 2 \rangle \subset \mathbb{Z}[x]$, explicitly?

EXAMPLE 1.15. (1) Let I and J be ideals of R. Then

$$I + J := \{a + b |\ a \in I, b \in J\}, \quad \text{and} \quad IJ := \{\sum_{finite} a_i b_i |\ a_i \in I, b_i \in J\}$$

are ideals of R, and $IJ \leq I \cap J$.

(2) For $a \in R$, $\operatorname{ann}(a) := \{r \in R |\ ra = 0\}$ is an ideal, which is called the annihilator of a.

(3) For an ideal I in R, $\operatorname{rad} I := \{r \in R |\ r^n \in I \text{ for some } n\}$ is an ideal, which is also denoted by $\sqrt{I}$.

(4) The set of all nilpotent elements in R forms an ideal, called the nilradical of R. We will see later that this is equal to the intersection of all prime ideals.

EXERCISE 1.16. Consider $R = M_{2\times 2}(\mathbb{R})$, and let $S = \left\{ \begin{pmatrix} a & b \\ 0 & d \end{pmatrix} |\ a, b, d \in \mathbb{R} \right\}$ and $I = \left\{ \begin{pmatrix} a & b \\ 0 & 0 \end{pmatrix} |\ a, b \in \mathbb{R} \right\}$.

(1) Show that S is a subring of R.

(2) Show that an ideal of R is 0 or R.

(3) Show that $S/I \cong \mathbb{R}$.

EXERCISE 1.17. Let R be an integral domain. Show that if R has a composition series then R is a field.
[Hint: If an integral domain is simple then it is a field. A field has a unit.]

We close this section with properties of ideals and quotient rings.

Theorem 1.18. *Let $\phi : R \to R'$ be a ring homomorphism.*
(a) *If $I \leq R$ is an ideal, then $\phi(I) \leq \phi(R)$ is an ideal.*
(b) *If $I' \leq \phi(R)$ is an ideal, then $\phi^{-1}(I') \leq R$ is an ideal.*

REMARK 1.19. In the above theorem, $\phi(I)$ need not be an ideal of the whole ring R'. For example, considering an embedding $\mathbb{Z} \hookrightarrow \mathbb{Q}$, $n\mathbb{Z}$ is not an ideal of $\mathbb{Q}$.

Theorem 1.20 (Isomorphism theorems).
(a) *Let $\phi : R \to S$ be a ring homomorphism. Then*

$$R/\ker\phi \cong \operatorname{Im}\phi.$$

(b) * *Let I and J be ideals of R. Then*

$$I/(I \cap J) \cong (I + J)/J.$$

(c) * *Let I and J be ideals of R such that $I \leq J$. Then $J/I \leq R/I$ and*

$$(R/I)/(J/I) \cong R/J.$$

Midterm Exam

(April 22, 2013)

1. (18 points) Determine whether each statement is true or not: True(T) or False(F). (If you answer correctly, you get 3 points; If nothing, 1 point; If wrong, 0 point.)
   1) The multiplication in $\mathbb{H} = \{\text{quaternions}\}$ is commutative.
   2) Every finite integral domain is a field.
   3) The set of zero divisors of $\mathbb{Z}_n$ with 1 is a group under multiplication.
   4) For a finite field $\mathbb{F}$, $\mathbb{F} \setminus \{0\}$ is a cyclic group under multiplication.
   5) For a prime p, $\Phi_p(x) = x^{p-1} + x^{p-2} + \cdots + x + 1$ is irreducible over $\mathbb{Q}$.
   6) The subring $\mathbb{Z}$ of $\mathbb{Q}$ is an ideal of $\mathbb{Q}$.

2. (12pts) Determine whether each of the following is easy(=efficient) or hard(=taking a long time) in the blank. (If you answer nothing, you get 1 point; If wrong, 0 point.)
   1) Gaussian elimination for a system of linear equations: (   )
   2) Euclidean algorithm for univariate polynomials: (   )
   3) Finding a generator for the multiplicative group of a finite field: (   )
   4) Factoring a univariate polynomial into the product of irreducibles: (   )

3. (10pts) Define the Boolean ring structure on $P(\{1,2,3\})$, that is, define two operations $+$ and $\times$. What is the identity under $+$, and $\times$, respectively? Is $P(\{1,2,3\})$ an integral domain?
4. (15pts) (1) Write down the definition of an integral domain.
   (2) State Fermat's little theorem.
   (3) State Eisenstein's criterion for the irreducibility of $f(x) \in \mathbb{Z}[x]$.

- For each of the following problems, you should justify your answer.

5. (10pts) Let $\mathbb{H}^\times$ be the multiplicative group of nonzero quaternions, and $GL_2(\mathbb{C})$ be group of invertible 2×2 complex matrices. If we define $\phi : \mathbb{H}^\times \to GL_2(\mathbb{C})$ by $\phi(\alpha + \beta j) = \begin{pmatrix} \alpha & \beta \\ -\bar{\beta} & \bar{\alpha} \end{pmatrix}$, then show that ϕ is a homomorphism under the multiplication.
6. (10pts) Find the unit group $\mathbb{Z}_{10}^\times$. Among $(\mathbb{Z}_4, +)$ and $(\mathbb{Z}_2 \times \mathbb{Z}_2, +)$, which is isomorphic to $\mathbb{Z}_{10}^\times$ as a group?

7. (10pts) Let $a(x) = x^4 - x^3 - 2x^2 + 6x - 4$ and $b(x) = x^3 - x^2 + 2$. Find the greatest common divisor $g(x) = \gcd(a(x), b(x))$, and also find polynomials $s(x)$ and $t(x)$ such that $a(x)s(x) + b(x)t(x) = g(x)$.

8. (15pts) Let D be an integral domain. Define an equivalence relation $\sim$ on $D \times (D \setminus \{0\})$:

$$(a, b) \sim (c, d) \text{ if and only if } ad - bc = 0.$$

We define $+$ on $F = \{[a, b] |\ a, b \in D, b \neq 0\}$ where $[a, b] = \{(x, y) |\ (x, y) \sim (a, b)\}$:

$$[a, b] + [c, d] := [ad + bc, bd].$$

Then show that this operation is well-defined.

2. Algebraic sets *

DEFINITION 2.1. (a) Let S be a subset of $\mathbb{C}[\mathbf{x}] = \mathbb{C}[x_1, \ldots, x_n]$. The *algebraic set* of S is defined by the common zero set

$$V(S) = \{a \in \mathbb{C}^n |\ f(a) = 0 \text{ for each } f \in S\}.$$

An algebraic set is called a *variety* if it is irreducible, i.e., it cannot be a union of two proper algebraic sets.

(b) Let $J \subseteq \mathbb{F}^n$. Then we define

$$I(J) = \{f \in \mathbb{C}[\mathbf{x}] |\ f(a) = 0 \text{ for each } a \in J\},$$

called the *vanishing ideal* of J.

DEFINITION 2.2. Let I be an ideal in a commutative ring R with 1. A subset $\{b_1, \ldots, b_r\}$ of I is a *basis*(a set of generators) if

$$I = \langle b_1, \ldots, b_r \rangle = \{\sum_{i=1}^{r} c_i b_i |\ c_i \in R \text{ for } 1 \leq i \leq r\}.$$

Proposition 2.3. *Let $S = \{f_1, \ldots, f_r\} \in \mathbb{C}[\mathbf{x}]$ and $I = \langle f_1, \ldots, f_r \rangle \leq \mathbb{C}[\mathbf{x}]$. Then*

$$V(S) = V(I).$$

Theorem 2.4 (Hilbert's basis theorem). *Every ideal in $\mathbb{F}[x_1, \ldots, x_n]$ has a finite basis.*

REMARK 2.5. In general, if a commutative ring R is Noetherian then $R[x]$ is Noetherian. We say that R is Noetherian if every ideal of R is finitely generated.

Proposition 2.6. *Let $f(\mathbf{x}), g(\mathbf{x}), q(\mathbf{x})$ and $r(\mathbf{x})$ be polynomials in $\mathbb{C}[\mathbf{x}]$ such that $f(\mathbf{x}) = g(\mathbf{x})q(\mathbf{x}) + r(\mathbf{x})$. Then*

$$V(f, g) = V(g, r).$$

Proof. Let $f, g \in B$, $g, r \in B'$ and $S = B \cup \{r\} = B' \cup \{f\}$. Then $B \subset S \subset \langle B \rangle$ and $B' \subset S \subset \langle B' \rangle$ because $r = f - gq \in \langle B \rangle$ and $f = gq + r \in \langle B' \rangle$. So $\langle B \rangle = \langle S \rangle = \langle B' \rangle$. □

EXAMPLE 2.7. Let $f(x) = x^4 + x^3 - 3x^2 - 5x - 2$ and $g(x) = x^3 + 3x^2 - 6x - 8 \in \mathbb{C}[x]$. Since $f(x) = g(x)(x-2) + 9x^2 - 9x - 18$ and $g(x) = (x^2 - x - 2)(x + 4)$, we get that $V(f, g) = V(g, x^2 - x - 2) = V(x^2 - x - 2, 0) = V(x^2 - x - 2) = \{-1, 2\} \subset \mathbb{C}$.

EXAMPLE 2.8. Consider an ideal $I = \langle xy^2, y^2 - y\rangle \le \mathbb{C}[x, y]$. A set $\{xy^2, y^2 - y\}$ is a basis of I. We divide xy^2 by $y^2 - y$ to have the remainder $xy^2 - (y^2 - y)x = xy$, so $I = \langle y^2 - y, xy\rangle$, that is, $\{y^2 - y, xy\}$ is a new basis of I.

Later in Chapter 8, we will thoroughly study a notion of a "good" basis of I and how to find it. Here, we briefly introduce what it is.

DEFINITION 2.9. We give a total ordering $<$, called a *monomial order*, on the set M of all monomials $x_1^{a_1}x_2^{a_2}\cdots x_n^{a_n} \in \mathbb{C}[\mathbf{x}]$ for all $a_i \in \mathbb{Z}_{\ge 0}$, satisfying:

1) $1 < m$ for all monomials $m \neq 1$;
2) For each $m_i, m_j \in M$, exactly one of $m_i < m_j, m_i = m_j, m_i > m_j$ holds;
3) $m_i < m_j$ and $m_j < m_k$ imply that $m_i < m_k$;
4) $m_i < m_j$ implies that $mm_i < mm_j$ for for all $m \in M$.

EXAMPLE 2.10. There are many examples of monomial orders. Typical ones are: lexicographic order, degree(graded, homogeneous) lex order, degree reverse lex order, weighted (reverse) lex order.

DEFINITION 2.11. A subset set $\{g_1, \ldots, g_r\}$ of $\mathbb{F}[x_1, \ldots, x_n]$ with term order $<$, is a *Gröbner basis* for the ideal $I = \langle g_1, \ldots, g_r\rangle$ if, for each nonzero $f \in I$, there exists g_i such that the leading monomial $\overline{g_i}$ of g_i divides the leading monomial $\overline{f}$ of f.

EXAMPLE 2.12. Let $I = \langle x^2y - 1, xy^2 - y\rangle \le \mathbb{C}[x, y]$. Then

$$xy - y = y(x^2y - 1) - x(xy^2 - y) \in I.$$

But, $x^2y \nmid xy$ and $xy^2 \nmid xy$, so $\{x^2y - 1, xy^2 - y\}$ is not a Gröbner basis of I.

DEFINITION 2.13. We multiply f and g by monomials as small as possible so that the leading monomials will be the same as $\operatorname{lcm}(\overline{f}, \overline{g})$, and then subtract or add with suitable coefficients from $\mathbb{F}$ so cancellation results. A polynomial in this fashion is denoted by $S(f, g)$, called the *S-polynomial* of f and g.

Theorem 2.14 (Buchberger 1965). [1] *A basis $G = \{g_1, \ldots, g_r\}$ is a Gröbner basis for the ideal $\langle g_1, \ldots, g_r\rangle$ if and only if for $i \neq j$, the polynomial $S(g_i, g_j)$ can be reduced to zero by repeatedly dividing remainders by elements of G.*

[1] Wolfgang Gröbner(1899–1980) was an Austrian mathematician. His student, Bruno Buchberger(1942–), introduced Gröbner bases in his Ph.D. thesis [B65]. Buchberger named them after his advisor [B76a].

Corollary 2.15 (Buchberger's algorithm). *We want to find a Gröbner basis from the given basis. First, we reduce the basis to the smallest degree by division, and choose arbitrary two polynomials in the basis. If $S(g_i, g_j)$ cannot be reduced to zero, we augment the given basis with this $S(g_i, g_j)$ and return to the first step. When every S-polynomial reduces to zero, we have arrived at a Gröbner basis. This process stops in a finite number of steps by Neotherian property.*

EXAMPLE 2.16. Let $I = \langle x^2y - 1, xy^2 - y\rangle \leq \mathbb{C}[x, y]$ with $G = \{x^2y - 1, xy^2 - y\}$. Then

$$S(x^2y - 1, xy^2 - y) = y(x^2y - 1) - x(xy^2 - y) = xy - y \in I.$$

Since $\langle x^2y - 1, xy^2 - y, xy - y\rangle = \langle y - 1, x - 1\rangle$ and $S(x - 1, y - 1) = x - y$ reduces to 0, we conclude that $\{x - 1, y - 1\}$ is a Gröbner basis of I.

3. Prime and maximal ideals

Let R be a ring. Note that $R/R \cong \{0\}$ and $R/\{0\} \cong R$. We say that R is an improper ideal and the zero ideal is the trivial ideal. An ideal I with $\{0\} \lneq I \lneq R$ is a nontrivial proper ideal of R.

REMARK 3.1. Let R be a ring with 1. An ideal I containing a unit u is the whole ring R, because $u \in I$ and $u^{-1} \in R$ implies that $1 = uu^{-1} \in I$ and, for each $r \in R$, $r = r \cdot 1 \in I$. So the ideals of of a field $\mathbb{F}$ are only $\{0\}$ and $\mathbb{F}$.

DEFINITION 3.2. Let R be a commutative ring. A proper ideal I of R is called a *prime ideal* of R if, for each $a, b \in R$ with $ab \in I$, we have $a \in I$ or $b \in I$. In general, for a ring R, an ideal I is prime if, each ideal A, B of R with $AB \subseteq I$ satisfy $A \subseteq I$ or $B \subseteq I$.

EXAMPLE 3.3. (1) $6\mathbb{Z}$ is not a prime ideal of $\mathbb{Z}$, since $2 \cdot 3 \in 6\mathbb{Z}$ but $2, 3 \notin 6\mathbb{Z}$. Note that $n\mathbb{Z}$ is a prime ideal if and only if n is a prime or 0.

(2) $\langle x^2 - 1\rangle$ is not a prime ideal of $\mathbb{Z}[x]$, since $(x + 1)(x - 1) \in \langle x^2 - 1\rangle$ but $x \pm 1 \notin \langle x^2 - 1\rangle$.

Theorem 3.4. *Let R be a commutative ring with* 1.

(a) *An ideal P is prime in R if and only if $R \setminus P$ is a multiplicative set.*

(b) *P is a prime ideal of R if and only if R/P is an integral domain.*

Proof. (b) A commutative ring R/I with $1 + I$ is an integral domain if it has no zero divisors, that is, $(a + I)(b + I) = I$ implies that $a + I = I$ or $b + I = I$. This condition is the same as that I is a prime ideal. □

Proposition 3.5. *Let $f : R \to S$ be a ring homomorphism with kernel K, I an ideal in R, J an ideal in S.*

1) *$f^{-1}(J)$ is an ideal in R that contains K, and $f(I)$ is an ideal in $f(S)$.*
2) *If f is an epimorphism, then*

$$\{\text{prime ideals in } R \text{ that contain } K\} \overset{1-1}{\longleftrightarrow} \{\text{prime ideals in } S\}$$

by $P \mapsto f(P)$. As a corollary, every prime ideal in R/I is of the form P/I, where P is a prime ideal in R that contains I.

EXERCISE 3.6. * (1) If every proper ideal of R is prime, then R is a field. [Hint: Suppose there exists a nonzero nonunit $a \in R$. Consider an ideal Ra^2.]

(2) The set of zero divisors of R is a union of prime ideals.

(3) Let M be a multiplicative subset of R not containing 0. If P is a maximal element in the set of ideals of R whose intersection with M is empty, then P is prime.

(4) Let $I \lneq R$ be an infinitely generated ideal such that any ideal properly containing I is finitely generated. Prove that I is a prime ideal.

Solution. (4) Suppose not, that is, there exist $x, y \notin I$ with $xy \in I$. Then $I \lneq I + Rx = \langle u_1, \ldots, u_n, x\rangle$ for $u_i \in I$. Let $(I : x) = \{r \in R |\ rx \in I\}$. Since $y \in (I : x) \gneq I$, $(I : x) = \langle v_1, \ldots, v_m\rangle$ for $v_i \in R$. Then $I = \langle u_1, \ldots, u_n, v_1x, \ldots, v_mx\rangle$, contradiction. □

DEFINITION 3.7. An ideal I of a ring R is called a *maximal ideal* of R if there is no ideal $J \lneq R$ properly containing I.

Theorem 3.8. *Let R be a commutative ring with* 1.

(a) *An ideal M is a maximal ideal of R if and only if R/M is a field.*

(b) *As a corollary, every maximal ideal in R is prime.*

Proof. (a) ⇒) In a commutative ring R/M with $1+M$, take a nonzero $a+M \in R/M$ (i.e., $a \notin M$). Consider

$$I = \{ra + m|\ r \in R, m \in M\},$$

which is an ideal properly containing M. (Check!) Then $I = R$, since M is maximal. In particular, $1 = ba + m \in I$ for some $b \in R$ and $m \in M$. That is,

$$1 + M = (ba + M) = (b + M)(a + M)$$

for some $b + M \in R/M$, which is $(a + M)^{-1} \in R/M$.

$\Leftarrow$) Suppose that there exists an ideal N such that $M \lneq N \lneq R$. Then, we consider the canonical projection $\gamma : R \to R/M$, thus $\{M\} \lneq \gamma(N) \lneq R/M$. This contradicts that a field R/M has no nontrivial proper ideals. □

Proposition 3.9. * *A proper ideal M in R is maximal if and only if, for each $r \in R \setminus M$, there exists $x \in R$ such that $1 - rx \in M$.*

EXAMPLE 3.10. (1) $\mathbb{Z} \times \{0\}$ is a prime ideal of $\mathbb{Z} \times \mathbb{Z}$, since the quotient is $\mathbb{Z}$, an integral domain. But, it is not a maximal ideal. Note that $n\mathbb{Z}$ is a maximal ideal of $\mathbb{Z}$ if and only if n is a prime.

(2) Consider $\langle x \rangle$ in $\mathbb{Z}[x]$. This is a prime ideal, but not a maximal ideal. Note that $\langle x \rangle < \langle x, 2 \rangle < \mathbb{Z}[x]$.

EXERCISE 3.11. Show that the ring $M_{n\times n}(D)$ (D, a division ring) has no nontrivial ideal. [Hint: Consider $E_{r,s}$, where $E_{r,s}$ has 1 as (r, s)-entry and 0 elsewhere.], and has zero divisors if $n \geq 2$. So 0 is a maximal and prime ideal in a noncommutative ring $M_{n\times n}(D)$.

Let R be a ring with 1 and define a map $\phi : \mathbb{Z} \to R$ by $n \mapsto n \cdot 1$. Then ϕ is a ring homomorphism. If $\operatorname{ch} R = n > 1$ then $R \geq \phi(\mathbb{Z}) \cong \mathbb{Z}/n\mathbb{Z}$. If $\operatorname{ch} R = 0$ then $R \geq \phi(\mathbb{Z}) \cong \mathbb{Z}$.

Proposition 3.12. *Let $\mathbb{F}$ be a field. Then we have two cases*:

(i) $\operatorname{ch} \mathbb{F} = p$ *(p, a prime) and* $\mathbb{F} \geq F' \cong \mathbb{Z}/p\mathbb{Z}$;
(ii) $\operatorname{ch} \mathbb{F} = 0$ *and* $\mathbb{F} \geq F' \cong \mathbb{Q}$.

Proof. (i) $F' \cong \mathbb{Z}_n$ where n must be a prime.

(ii) Since $\mathbb{F}$ contains $D \cong \mathbb{Z}$, it contains the field F', isomorphic to the field $\mathbb{Q}$ of quotients of $\mathbb{Z}$. □

DEFINITION 3.13. The fields $\mathbb{Z}_p$ (p, a prime) and $\mathbb{Q}$ are called *prime fields.*

4. Ideal structure of $\mathbb{F}[x]$

Now we study the ideal structure in the polynomial ring $\mathbb{F}[x]$ over a field $\mathbb{F}$. Recall that an ideal of R is principal if it is generated by a single element of R. Let a be an element in a commutative ring R with 1. Then we write

$$\langle a \rangle = \{ra \mid r \in R\}.$$

Theorem 4.1. *Every ideal in the polynomial ring $\mathbb{F}[x]$ over a field $\mathbb{F}$ is principal.*

Proof. Let I be an ideal of $\mathbb{F}[x]$. If $I = \{0\}$ then $I = \langle 0 \rangle$. Assume that $I \neq \{0\}$. Then we take $g(x)$ to be any nonzero element of I of minimal degree. If $\deg g = 0$, i.e., $g \in \mathbb{F} \setminus \{0\}$, since g is a unit in $\mathbb{F}[x]$, we have $I = \mathbb{F}[x] = \langle 1 \rangle$.

If $\deg g(x) \geq 1$, then, for each $f(x) \in I$, we divide by $g(x)$ to get that

$$f(x) = g(x)q(x) + r(x)$$

with $\deg r < \deg g$. Since $r(x) = f(x) - g(x)q(x) \in I$ and $\deg r < \deg g$, $r(x)$ must be 0, thus $f(x) = g(x)q(x)$, that is, $I = \langle g(x) \rangle$. □

Remark 4.2. In the above theorem, the condition that $\mathbb{F}$ is a field is essential. Note that every ideal of $\mathbb{Z}[x]$ is not principal.

Example 4.3. In any commutative ring R, the ideal generated by $a_1, a_2, \ldots, a_n$ is

$$\langle a_1, a_2, \ldots, a_n \rangle = \{\sum_{i=1}^{n} r_i a_i \mid r_i \in R\}.$$

Note that, in $\mathbb{Q}[x]$,

$$\langle x^3 + 3x^2 - 4, x^2 + 2x - 3, (x-1)^2 \rangle = \langle x - 1 \rangle,$$

which is principal.

Remark 4.4. * Ideals in the polynomial ring $\mathbb{F}[x, y]$ are not principal any more. Therefore, the computations in the multivariate polynomial rings become more complicated. We will study basic theory in Chapter 8. Note that, in $\mathbb{Q}[x, y]$,

$$\langle xy^2 - x, xy - x, y^2 - x \rangle = \langle xy - x, y^2 - x \rangle = \langle x^2 - x, xy - x, y^2 - x \rangle,$$

which is never principal.

Theorem 4.5. *Let $p(x) \in \mathbb{F}[x]$. Then $\langle p(x) \rangle$ is a maximal ideal if and only if $p(x)$ is irreducible over $\mathbb{F}$.*

Proof. $\Rightarrow$) Let $p(x) = f(x)g(x)$ in $\mathbb{F}[x]$ with $\deg p(x) = n \geq 1$. Since a maximal ideal is prime, $p(x) = f(x)g(x) \in \langle p(x)\rangle$ implies that $f(x) \in \langle p(x)\rangle$ or $g(x) \in \langle p(x)\rangle$. Thus it is impossible that $\deg f(x) < n$ and $\deg g(x) < n$.

$\Leftarrow$) Suppose that there exists an ideal I of $\mathbb{F}[x]$ such that $\langle p(x)\rangle \leq I \leq \mathbb{F}[x]$. Since I is principal, $I = \langle g(x)\rangle$ for some $g(x) \in I$. From $p(x) \in I$, we have $p(x) = g(x)q(x)$ for some $q(x) \in \mathbb{F}[x]$. Then, the irreducibility of $p(x)$ says that $\deg g = 0$ or $\deg q = 0$, that is, $I = \mathbb{F}[x]$ or $I = \langle p(x)\rangle$, which proves that $\langle p(x)\rangle$ is a maximal ideal. $\square$

EXAMPLE 4.6. (1) The quotient ring $\mathbb{Q}[x]/\langle x^2 - 2\rangle$ is a field.

(2) The ideal $\langle x^2 + 1\rangle$ in $\mathbb{R}[x]$ is maximal. However $\langle x^2 + 1\rangle \trianglelefteq \mathbb{C}[x]$ is not prime, so it is not maximal

(3) Consider $f(x) = x^3 + 3x + 2$. Note that this is irreducible in $\mathbb{Z}_5[x]$, since none of $f(0), f(\pm 1), f(\pm 2)$ is zero. So $\langle x^3 + 3x + 2\rangle$ is a maximal ideal of $\mathbb{Z}_5[x]$ and the quotient ring $\mathbb{Z}_5[x]/\langle x^3 + 3x + 2\rangle$ is a field.

EXERCISE 4.7. (1) What is the multiplicative inverse of $x + \langle x^2 - 2\rangle$ in $\mathbb{Q}[x]/\langle x^2 - 2\rangle$?

(2) Find all maximal ideals of $\mathbb{C}[x]$, where $\mathbb{C}$ is the field of complex numbers.

(3) Let $\mathbb{F}$ be a field. Show that every proper nontrivial prime ideal of $\mathbb{F}[x]$ is maximal.

REMARK 4.8. If we have an irreducible polynomial $p(x)$ over a field $\mathbb{F}$, then we obtain an embedding

$$\psi : \mathbb{F} \hookrightarrow \mathbb{F}[x]/\langle p(x)\rangle$$

by $\psi(a) = a + \langle p(x)\rangle$. Note that the field $\mathbb{F}[x]/\langle p(x)\rangle$ is an extension field of $\mathbb{F}$ containing a zero of $p(x)$, since $\phi_\alpha(p(x)) = 0$ where ϕ_α is the evaluation homomorphism at $\alpha = x + \langle p(x)\rangle$. This is a basic goal in Chapter 6, due to Kronecker[2].

Theorem 4.9. *Let $p(x)$ be an irreducible polynomial in $\mathbb{F}[x]$. Then $p(x)|r(x)s(x)$ for each $r(x), s(x) \in \mathbb{F}[x]$ implies that $p(x)|r(x)$ or $p(x)|s(x)$.*

Proof. Note that a maximal ideal $\langle p(x)\rangle$ is a prime ideal. So $r(x)s(x) \in \langle p(x)\rangle$ implies that $r(x) \in \langle p(x)\rangle$ or $s(x) \in \langle p(x)\rangle$. $\square$

EXERCISE 4.10. * What is a prime ideal of $\mathbb{F}[x, y]$?

[2]Leopold Kronecker(1823–1891), a German mathematician. He said, "God made natural numbers; all else is the work of man."

In the remaining of this section, we treat more advanced materials concerning the prime and maximal ideals.

Theorem 4.11. * *An (affine) algebraic set $V \subset \mathbb{C}^n$ is irreducible if and only if the defining ideal $I(V)$ is prime.*

EXAMPLE 4.12. Let $f_1 = xy^2 + y^2 + x + 1$, $f_2 = y^2 - 1$, and $I = \langle f_1, f_2 \rangle \in \mathbb{C}[x, y]$. Determine if I is prime.

Solution. Since $V(I)$ consists of 2 points, the ideal I cannot be prime.

Another method is using Gröbner bases. Fix the lex order with $x > y$. A Gröbner basis for I with respect to $>$ is $G = \{g_1 = y^2 - 1, g_2 = x + 1\}$. Thus, the remainder by division of $f \in \mathbb{C}[x, y]$ is of the form $ay + b (a, b \in \mathbb{C})$, and $\mathbb{C}[x, y]/I$ is a $\mathbb{C}$-vector space with a basis $\{1, y\}$. Its ring structure is determined by the multiplication table. We easily compute to obtain that $y \pm 1$ are (the only) zero divisors of the ring $\mathbb{C}[x, y]/I$. In this zero-dimensional example,

$$|V(I)| = |\{(-1, 1), (-1, -1)\}| = 2 = \dim_{\mathbb{C}} \mathbb{C}[x, y]/I.$$

In general, this is true if I is a radical ideal. □

Theorem 4.13 (Prime avoidance). * *Let $I, P_1, \ldots, P_r$ be ideals in R. Suppose that all but possibly two of the P_i's are prime ideals. Then, if $I \not\subseteq P_i$ for each i, the ideal I is not contained in the set-theoretical union $\bigcup_i P_i$.*

Proof. [M, pp.2–3] Omitting those P_i such that $P_i \subseteq P_j$ for some j, we may assume that there are no inclusion between the P_i's. We use induction on r. When $r = 2$, suppose $I \subseteq P_1 \cup P_2$. Take $x \in I \setminus P_2$ and $s \in I \setminus P_1$. Then $x \in P_1$, hence $s + x \notin P_1$, therefore both s and $s + x$ must be in P_2. Then $x \in P_2$, contradiction.

When $r > 2$, assume that P_r is prime. Then $IP_1 \cdots P_{r-1} \not\subseteq P_r$; Take an element $x \in IP_1 \cdots P_{r-1} \setminus P_r$. Put $S = I \setminus (P_1 \cup \cdots \cup P_{r-1})$. By induction hypothesis, $S \neq \emptyset$. Suppose $I \subseteq P_1 \cup \cdots \cup P_r$. Then $S \subseteq P_r$. But if $s \in S$ then $s + x \in S$ and therefore both s and $s + x$ are in P_r, hence $x \in P_r$, contradiction. □

DEFINITION 4.14. * The *spectrum* of a ring R is defined by, as a set,

$$\mathrm{Spec} R := \{\text{all prime ideals of } R\}.$$

Let $\mathbb{A} := k^n$, and $A := k[x_1, \ldots, x_n]$. A map $\varphi : \mathbb{A} \hookrightarrow \mathrm{Spec} A$ by

$$(a_1, \ldots, a_n) \mapsto \langle x_1 - a_1, \ldots, x_n - a_n \rangle$$

gives an embedding of $\mathbb{A}$ into Spec$A$. Note that Spec$A$ is topologized as follows[3]: For any ideal I of A, put

$$V(I) = \{P \in \text{Spec}A \mid I \subseteq P\},$$

and take as the closed sets in Spec$A$ all subsets of the form $V(I)$.

Theorem 4.15 (Chinese remainder theorem). * *Let $I_1, \ldots, I_n$ be ideals in R such that $I_i + I_j = R$ for all $i \neq j$.*

(a) *Given elements $x_1, \ldots, x_n \in R$, there exists $x \in R$ such that $x \equiv x_i \pmod{I_i}$ for $\forall\, i$. Furthermore, x is uniquely determined up to congruence modulo the ideal $I_1 \cap \cdots \cap I_n$.*

(b) *$I_1 \cap \cdots \cap I_n = I_1 I_2 \cdots I_n$ and $R/\bigcap I_i \cong \prod(R/I_i)$.*

EXERCISE 4.16. For any distinct maximal ideals $M_1, \ldots, M_n$, the above (a) and (b) hold.

Solution. First, we show the case of $n = 2$. Clearly, $M_1M_2 \subseteq M_1 \cap M_2$. For $x \in M_1 \cap M_2$, there exist $b_i \in M_i$ $(i = 1, 2)$ such that $b_1 + b_2 = 1$ ($\because$ $M_1 + M_2 = R$), i.e., $x = x \cdot 1 = x(b_1 + b_2) = xb_1 + xb_2 \in M_1M_2$, so $M_1 \cap M_2 \subseteq M_1M_2$, we have $M_1 \cap M_2 = M_1M_2$.

We show that the canonical homomorphism $\phi : R/(M_1 \cap M_2) \to R/M_1 \times R/M_2$ is an isomorphism. ϕ is clearly 1-1. For $(x_1 + M_1, x_2 + M_2) \in R/M_1 \times R/M_2$, there exist $b_i \in M_i$ $(i = 1, 2)$ such that $x_1 - x_2 = b_1 - b_2 \in M_1 + M_2 = R$. Let $x = x_1 - b_1 = x_2 - b_2$. Then $x \equiv x_1 \pmod{M_1}$ and $x \equiv x_2 \pmod{M_2}$, i.e., ϕ is surjective, so ϕ is an isomorphism.

For the case of $n > 2$, since $M_1M_2 + M_3 = R$ ($\because$ For distinct maximal ideals M_i's, $M_1M_2 \not\subseteq M_3$.) and inductively $M_1 \cdots M_{n-1} + M_n = R$, we have $M_1 \cap \cdots \cap M_n = M_1 \cdots M_n$ and $R/\bigcap M_i \cong \prod(R/M_i)$. □

EXERCISE 4.17. * Let R be an integral domain and I be an ideal which is a product of distinct maximal ideals in two ways, say $I = m_1 \cdots m_r = n_1 \cdots n_s$. Prove that two factorizations are the same, except for the ordering of the terms.

---

[3]This topology is called the *Zariski topology*.

Final Exam

(June 17, 2011)

1. (30 points) Determine whether each statement is true or not: True(T) or False(F). (If you answer correctly, you get 3 points; If nothing, 1 point; If wrong, 0 point.)
   1) Every subring of a ring R is an ideal of R.
   2) Any subring $n\mathbb{Z}$ $(n = 0, 1, 2, \ldots)$ of $\mathbb{Z}$ is an ideal of $\mathbb{Z}$.
   3) The subring $\mathbb{Z}$ of $\mathbb{Q}$ is an ideal of $\mathbb{Q}$.
   4) The subring $\mathbb{Q}$ of $\mathbb{R}$ is an ideal of $\mathbb{R}$.
   5) A field has a nontrivial proper ideal.
   6) A field is an integral domain.
   7) Every prime ideal of a commutative ring R with 1 is a maximal ideal of R.
   8) $\langle x \rangle \leq \mathbb{Z}[x]$ is a maximal ideal of $\mathbb{Z}[x]$.
   9) $\langle p(x) \rangle \leq \mathbb{R}[x]$ is a maximal ideal of $\mathbb{R}[x]$ if and only if $p(x)$ is irreducible over $\mathbb{R}$.
   10) The field $\mathbb{Q}$ does not have a proper subring which is a field.
2. (10pts) Find all possible inclusions among the following:
   integral domain, commutative ring, field, division ring, skew-field.

- For each of the following problems, you should justify your answer.

3. (20pts) (1) Write down the definition of an ideal I in a ring R.
   (2) Write down the definition of a prime ideal P in a commutative ring R.
   (3) Find all prime ideals of $\mathbb{Z}$.
4. (20pts) Let $\mathbb{C}$ be the field of complex numbers.
   (1) Find the kernel of a ring homomorphism $\phi : \mathbb{R}[x] \to \mathbb{C}$ defined by $\phi(1) = 1$ and $\phi(x) = i$.
   (2) Using the fundamental homomorphism theorem in (1), which ring is isomorphic to $\mathbb{C}$?
5. (20pts) (1) Find the multiplicative inverse of $x + \langle x^2 + 1 \rangle$ in $\mathbb{R}[x]/\langle x^2 + 1 \rangle$.
   (2) Find the multiplicative inverse of $x + \langle x^2 + x + 1 \rangle$ in $\mathbb{R}[x]/\langle x^2 + x + 1 \rangle$.

Midterm Exam

(April 22, 2010)

1. (15 points) Determine whether each statement is true or not: True(T) or False(F). (If you answer correctly, you get 3 points; If nothing, 1 point; If wrong, 0 point.)
   1) The set $\mathbb{Z}_p = \{0, 1, 2, \ldots, p-1\}$ (p, a prime) with the addition and multiplication (mod p) is a field.
   2) Any subring of a ring R is an ideal of R.
   3) Every prime ideal of a ring is a maximal ideal.
   4) Let $\mathbb{F}[x]$ be the ring of polynomials with coefficients in a field $\mathbb{F}$. The number of zeros of $f(x) \in \mathbb{F}[x]$ is less than or equal to the degree $\deg f(x)$.
   5) Every ideal in $\mathbb{F}[x]$ is principal, that is, any ideal of $\mathbb{F}[x]$ is generated by one polynomial.

2. (30 pts) (1) Use Fermat's little theorem to find all zeros in $\mathbb{Z}_5$ of $2x^{219} + 3x^{74} + 2x^{57} + 3x^{44}$.
   (2) Describe all solutions of $5x \equiv 9 \pmod 6$.
   (3) Describe all solutions of $45x \equiv 15 \pmod{24}$.
3. (25 pts) (1) Show that $x^2 + x + 1$ is irreducible in $\mathbb{Q}[x]$.
   (2) Find the multiplicative inverse of $x + \langle x^2+x+1\rangle$ in the field $\mathbb{Q}[x]/\langle x^2+x+1\rangle$.
4. (30 pts) Let R be a commutative ring with 1 and I be a subring.
   (1) Write down the definition of an ideal I of R.
   (2) Write down the definition of a prime ideal I of R.
   (3) Find all prime ideals of the ring $\mathbb{Z}$ of integers.

5. Zorn's lemma *

DEFINITION 5.1. (a) The pair $(S, \leq)$ is a partially ordered set(poset) if the binary relation $\leq$ on S is reflexive, antisymmetric and transitive on S.

(b) A subset T of a poset $(S, \leq)$ is a chain if any two elements in T are comparable.

Zorn's lemma:[4] For any nonempty partially ordered set $(S, \leq)$, if every chain in S has an upper bound, then there is a maximal element in S.

FACT 5.2. The following are equivalent:

(1) Zorn's lemma; (2) Axiom of choice;

(3) Well-ordering principle; (4) Hausdorff[5] maximum principle.

EXAMPLE 5.3. (1) Every nonzero vector space has a basis.

(2) Every proper ideal in a ring with 1 is contained in a maximal ideal.

(3) Every field $\mathbb{F}$ has an algebraic closure $\overline{\mathbb{F}}$.

(4) Every field extension possesses a transcendence base.

(5) Every submodule E of a free module F over a principal ideal domain R is free. In fact, F has an R-basis $\{v_i\}_{i\in I}$, then E has a basis indexed by a subset of I.

(6) An R-module E is called *semisimple* if the following equivalent conditions hold:[6]

1) E is the sum of simple submodules.

2) E is the direct sum of simple submodules.

3) Every submodule F of E is a direct summand, i.e., there exists a submodule F' such that $E = F \oplus F'$.

(7)*(Tychonoff[7]) An arbitrary product of compact topological spaces is compact.

---

[4] Max August Zorn(1906–1993), a German mathematician. Zorn's lemma was first postulated by Kazimierz Kuratowski(1896–1980, Polish mathematician) in 1922, and then independently by Zorn in 1935.

[5] Felix Hausdorff(1868–1942), a German mathematician. He is one of the founders of modern topology. It is noticeable that there is a piano solo 'Hausdorff space' composed by Chunggil Kim(1933–2012) in 1975. Consult `http://www.youtube.com/watch?v=dpkYqYOH2MY`.

[6] Refer to [L, Ch.XVII §2].

[7] Andrey Nikolayevich Tikhonov(1906–1993), a Soviet and Russian mathematician. He gave contributions to topology, functional analysis, mathematical physics, and ill-posed problems.

(8)*(Poincaré[8] duality) For an orientable n-dimensional compact manifold M, there is an isomorphism between its cohomology and homology, i.e.,

$$H^q(M) \cong H_{n-q}(M).$$

Proof. (2) Let I be a proper ideal of R. We consider $\Sigma = \{J \subsetneq R \mid J \text{ is an ideal containing } I\}$, which is a poset by set-theoretic inclusion. Σ is not empty ($\because I \in \Sigma$), and every chain $\{C_i\}$ has an upper bound $\bigcup C_i$ in Σ ($\because \bigcup C_i$ is an ideal and $1 \notin \bigcup C_i$). So we apply Zorn's lemma to obtain that Σ has a maximal element, i.e., R has a maximal ideal containing I. □

REMARK 5.4. Be cautious that the condition in (2) that a ring has 1 is necessary. Show that $x\mathbb{R}[x]$ has no maximal ideal. [Hint: Consider an ideal $x^2\mathbb{R}[x]$ and use that a divisible abelian group has no maximal subgroup.]

6. Divisibility in integral domains

We know very well the structure of integral domains $\mathbb{Z}$ and $\mathbb{F}[x]$. In this section, we study the property of divisibility in an integral domain, and show that

$$ED \Longrightarrow PID \Longrightarrow UFD.$$

DEFINITION 6.1. Let D be an integral domain.

(a) An element $a \in D$ divides $b \in D$, or a is a factor of b if there exists $c \in D$ such that $b = ac$.

(b) An element $u \in D$ is a unit if u divides 1. Two elements a and b in D are associates if $a = bu$ for a unit $u \in D$.

(c) An element $p \in D$ is an *irreducible* if p is a nonzero nonunit, and $p = ab$ for $a, b \in D$ implies that a or b is a unit.

EXAMPLE 6.2. In $\mathbb{Z}$, the only units are ± 1. So the associates of 26 are ± 26. The prime numbers are irreducibles in $\mathbb{Z}$.

DEFINITION 6.3. An integral domain D is called a *principal ideal domain*(PID) if any ideal of D is principal, that is, any ideal is generated by one element.

[8] Jules Henri Poincaré(1854–1912), a French mathematician, theoretical physicist, engineer, and a philosopher of science.

Lemma 6.4. *Let D be a PID. If $I_i \leq D$ be ideals such that $I_1 \subseteq I_2 \subseteq \cdots$, then there exists a positive integer r such that $I_r = I_s$ for all $s \geq r$. That is, every chain of ideals is stationary. We say that every PID satisfies the* ascending chain condition*(ACC) for ideals.*

Proof. Let $I_1 \subseteq I_2 \subseteq \cdots$ be an ascending chain of ideals of a PID D. Consider $I = \bigcup_{i=1}^{\infty} I_i$, which is an ideal of D. Thus there exists $c \in D$ such that $I = \langle c \rangle$, so $c \in I_r$ for a positive integer r, which means that $\langle c \rangle \subseteq I_r \subseteq I_s \subseteq I = \langle c \rangle$. Hence we proved that $I_r = I_s$ for all $s \geq r$. □

EXAMPLE 6.5. The integral domain $\mathbb{Z}$ satisfies the ascending chain condition for ideals. However, $\mathbb{Z}$ does not satisfy the descending chain condition, since

$$\mathbb{Z} \gneq 2\mathbb{Z} \gneq 4\mathbb{Z} \gneq \cdots \gneq 2^n\mathbb{Z} \gneq 2^{n+1}\mathbb{Z} \gneq \cdots.$$

Lemma 6.6. *Let D be an integral domain and $a, b \in D$. Then $\langle a \rangle \subseteq \langle b \rangle$ if and only if b divides a. The equality $\langle a \rangle = \langle b \rangle$ means that a and b are associates.*

EXERCISE 6.7. Let R be a PID, S an integral domain, and $f : R \to S$ an epimorphism. Show that if f is not a monomorphism then S is a field.
[Hint: $0 \lneq f^{-1}(0) = \langle p \rangle$ for some irreducible $p \in R$.]

Theorem 6.8. *Let D be a PID. Then every nonzero nonunit element in D is a finite product of irreducibles.*

Proof. Let $a \neq 0$ be a nonunit in D. If a is not an irreducible, then $a = a_1 b_1$ where a_1 and b_1 are not units in D. So $\langle a \rangle \subsetneq \langle a_1 \rangle$. Repeatedly, we have a strictly ascending chain of principal ideals $\langle a \rangle \subsetneq \langle a_1 \rangle \subsetneq \langle a_2 \rangle \subsetneq \cdots$. By ACC for ideals in a PID, this chain is of finite length with the last term $\langle a_r \rangle$. Thus, we get that $a_r \in D$ is an irreducible, that is, a has an irreducible factor a_r.

Let $a = p_1 c_1$ for an irreducible p_1. If c_1 is not an irreducible, the same argument produces that $c_1 = p_2 c_2$ for for an irreducible p_2. In this process, we have a chain of ideals $\langle a \rangle \subsetneq \langle c_1 \rangle \subsetneq \cdots \subsetneq \langle c_{n-1} \rangle \subsetneq \langle 1 \rangle = D$ for some n, and conclude that $a = p_1 p_2 \cdots p_n$ for irreducibles $p_1, \ldots, p_n$. □

The following theorem is a generalization of Theorem 4.5 for $D = \mathbb{F}[x]$.

Theorem 6.9. *Let D be a PID which is not a field, and $\langle p \rangle \leq D$ be an ideal. Then $\langle p \rangle$ is maximal if and only if p is an irreducible.*

Proof. Let $p = ab$ in D, i.e., $\langle p \rangle \subseteq \langle a \rangle$.

$\Rightarrow$) If $\langle p \rangle = \langle a \rangle$, a and p are associates and b is a unit. If $\langle p \rangle \subsetneq \langle a \rangle$, then $\langle a \rangle = D$ since $\langle p \rangle$ is maximal, so a is a unit. Hence, p is an irreducible in D.

$\Leftarrow$) If a is a unit, then $\langle a \rangle = \langle 1 \rangle = D$. If b is a unit, then a and p are associates and $\langle p \rangle = \langle a \rangle$. Hence, $\langle p \rangle$ is a maximal ideal of a PID D. □

Proposition 6.10. *Let D be a PID and $p, a, b \in D$ with p, irreducible. If $p|ab$, then $p|a$ or $p|b$.*

Proof. By the above theorem, $\langle p \rangle$ is a maximal ideal of D. Thus $\langle p \rangle$ is a prime ideal of D by Theorem 3.8. So $ab \in \langle p \rangle$ implies that $a \in \langle p \rangle$ or $b \in \langle p \rangle$. □

Corollary 6.11. *Let $p, a_1, a_2, \ldots, a_n$ be elements in a PID D. If p is an irreducible and $p|a_1 a_2 \cdots a_n$, then $p|a_i$ for at least one i.*

DEFINITION 6.12. A nonzero nonunit element p in an integral domain D is called a *prime* in D if $p|ab$ for any $a, b \in D$ implies that $p|a$ or $p|b$.

EXAMPLE 6.13. Consider an ideal $I = \langle x^3, xy, y^3 \rangle \leq \mathbb{F}[x, y]$. The element xy is not a prime in I, since $xy|x^3y^3$ but $xy \nmid x^3$ and $xy \nmid y^3$. Also, neither x^3 nor y^3 is a prime.

REMARK 6.14. Any prime $p \in D$ is an irreducible in D. The converse is true if D is a unique factorization domain.

Proposition 6.15. *Let D be an integral domain. The following are equivalent*:

(i) *For any nonzero, nonunit element $a \in D$, the process of factoring a into irreducible elements of D terminates in finitely many steps*;

(ii) *D satisfies the ascending chain condition for principal ideals.*

EXAMPLE 6.16. $\mathbb{Z}[\sqrt{2}, \sqrt[4]{2}, \ldots, \sqrt[2^n]{2}, \ldots]$ is not a factorization domain.

DEFINITION 6.17. An integral domain D is called a *unique factorization domain*(UFD) if each nonzero nonunit element in D is written as a finite product of irreducible elements, uniquely up to the order of irreducibles and up to multiples of units.

Proposition 6.18. *Let D be an integral domain in which there is a factorization for each nonzero nonunit element. Then D is a UFD if and only if every irreducible element is a prime.*

EXAMPLE 6.19. * There are many integral domains that are not UFD.

(1) $\mathbb{F}[x, y]/\langle x^2 - y^2 - 1\rangle$ [$\because y^2$ factors as yy and as $(x+1)(x-1)$.],

(2) $\mathbb{Z}[\sqrt{-3}] = \{a + b\sqrt{-3} |\ a, b \in \mathbb{Z}\}$ [$\because 4 = 2 \cdot 2 = (1+\sqrt{-3})(1-\sqrt{-3})$.],

(3) $\mathbb{Z}[\sqrt{-5}] = \{a + b\sqrt{-5} |\ a, b \in \mathbb{Z}\}$ [$\because 6 = 2 \cdot 3 = (1+\sqrt{-5})(1-\sqrt{-5})$.],

(4) $\mathbb{Z}[\sqrt{10}] = \{a + b\sqrt{10} |\ a, b \in \mathbb{Z}\}$ [$\because 6 = 2 \cdot 3 = (4+\sqrt{10})(4-\sqrt{10})$.],

(5) $\mathbb{R}[\sin x, \cos x]$ [$\because \sin^2 x = (1+\cos x)(1-\cos x)$.][9].

Theorem 6.20. *Every PID is a UFD.*

Proof. Due to Theorem 6.8, we have only to show the uniqueness of factorization.

Let $a = p_1p_2\cdots p_r = q_1q_2\cdots q_s$ where p_i's and q_j's are irreducibles. By changing the order if necessary, $p_1|q_1q_2\cdots q_s$ implies that $p_1|q_1$ so $q_1 = p_1u_1$ for some unit u_1. The cancellation law in D leads us that $p_2\cdots p_r = u_1q_2\cdots q_s$. Continuing this process, we get that $1 = u_1\cdots u_rq_{r+1}\cdots q_s$, hence we must have $r = s$. □

EXAMPLE 6.21. (1) $\mathbb{Z}$ and $\mathbb{F}[x]$ are UFD.

(2)* $\mathbb{F}[[x]] = \{\sum_{i=1}^{\infty} a_ix^i |\ a_i \in \mathbb{F}\}$ is a UFD. In fact, $\mathbb{F}[[x]]$ is a UFD with a unique irreducible element, or a local PID not a field, or a local Dedekind domain, i.e., a *discrete valuation ring.*

EXERCISE 6.22. (1)* An integral domain D is a UFD if and only if every nonzero prime ideal in D contains a nonzero principal ideal that is prime.

(2) If D is a UFD and $a_1, \ldots, a_n$ are elements of D, then there exists a great common divisor of $a_1, \ldots, a_n$. If D is a PID, then $\gcd(a_1, \ldots, a_n)$ is of the form $r_1a_1 + \cdots + r_na_n$ $(r_i \in D)$.

Now we consider some facts concerning polynomials over a UFD.

DEFINITION 6.23. Let D be a UFD. A polynomial $f(x) = a_0 + a_1x + \cdots + a_nx^n \in D[x]$ is primitive if $\gcd(a_0, a_1, \ldots, a_n)$ is a unit in D.

EXAMPLE 6.24. $4x^2 + 6x + 2 = 2(2x^2 + 3x + 1) \in \mathbb{Z}[x]$. $2x^2 + 3x + 1$ is primitive in $\mathbb{Z}[x]$.

Lemma 6.25. *Let D be a UFD.*

[9]In the ring of trigonometric polynomials over $\mathbb{R}$, the units are the constant functions and all elements of degree 1 are irreducible. $\mathbb{C}[\sin x, \cos x]$ is the same as $\mathbb{C}[z, z^{-1}]$ where $z = e^{it}$. For details, see H.F.Trotter, *An Overlooked Example of Nonunique Factorization*, The American Mathematical Monthly, 1988, pp.339–342.

(a) *For a nonconstant polynomial* $f(x) \in D[x]$, *there exist* $c \in D$ *and primitive* $g(x) \in D[x]$ *such that* $f(x) = cg(x)$. *A content* c *of* $f(x)$ *and a primitive element* $g(x)$ *are unique up to a unit factor in* D.

(b) (Gauss) *A product of two primitive polynomials in* $D[x]$ *is again primitive.*

(c) *Let* F *be the quotient field of* D *and* $f \in D[x]$ *with positive degree. If* f *is irreducible in* $D[x]$, *then* f *is also irreducible in* $F[x]$. *Conversely, if* f *is primitive in* $D[x]$ *and irreducible in* $F[x]$, *then* f *is irreducible in* $D[x]$.

Proof. (a) Let c be a greatest common divisor among the coefficients of $f(x)$. We write $f(x) = cg(x) = dh(x)$ with g, h primitive. Then $c|d$ and $d|c$, that is, c and d are associates.

(b) Let $f(x) = a_0 + a_1x + \cdots + a_nx^n$ and $g(x) = b_0 + b_1x + \cdots + b_mx^m$. For each irreducible p satisfying that $p|a_i$ for $i < r$ and $p \nmid a_r$, and $p|b_j$ for $j < s$ and $p \nmid b_s$, we consider the polynomial $f(x)g(x) = \sum_{i=0}^{n+m} c_ix^i$, where

$$c_{r+s} = (a_0b_{r+s} + \cdots + a_{r-1}b_{s+1}) + a_rb_s + (a_{r+1}b_{s-1} + \cdots + a_{r+s}b_0).$$

Then $p \nmid a_rb_s$ and $p \nmid c_{r+s}$. Thus, $f(x)g(x)$ is primitive.

(c) The first statement follows from the contrapositive, and the converse is clear from $D[x] \subseteq F[x]$. □

The following is a generalization of Ch.4 Lemma 5.25.

Lemma 6.26 (Eisenstein's criterion). *Let* D *be a UFD with quotient field* F *and* $p \in D$ *be a prime. Suppose that* $f(x) = a_nx^n + \cdots + a_0 \in D[x]$, *and* $a_n \not\equiv 0 \pmod{p}$, *but* $a_i \equiv 0 \pmod{p}$ *for* $i < n$, *with* $a_0 \not\equiv 0 \pmod{p^2}$. *Then* $f(x)$ *is irreducible over* F.

EXERCISE 6.27. 1. Let D be an integral domain and $f(x) \in D[x]$.

(1) Show that $f(x)$ is irreducible in $D[x]$ if and only if $f(x-c)$ is irreducible $D[x]$ $(c \in D)$.

(2) Show that for each prime p, the cyclotomic polynomial $f(x) = 1+x+\cdots x^{p-2}+x^{p-1}$ is irreducible in $\mathbb{Z}[x]$.

2. Let $\mathbb{Z}_m$ be the ring of integers modulo m.

(1) Show that $x^3 + 2x + 1$ is irreducible in $\mathbb{Z}_m[x]$.

(2) Show that the ideal I generated by $\{5, x^3 + 2x + 1\}$ in $\mathbb{Z}[x]$ is maximal.

[Hint: $(Ra + Rb)/Ra \cong Rb/(Ra \cap Rb) \cong (R/Ra)b$.]

3. Find all prime ideals of $\mathbb{C}[x]$ and $\mathbb{C}[[x]]$.

In the following theorem, we prove an important fact that the polynomial ring over a UFD is again a UFD.

Theorem 6.28. *If D is a UFD, then $D[x]$ is a UFD.*

Proof. Let $f(x) \in D[x] \backslash D$, that is, $\deg f(x) \geq 1$. Since $D[x] \subseteq F[x]$ (F, the quotient field of D) and $F[x]$ is a UFD, we factorize $f(x)$ in $F[x]$, say, $f(x) = p_1(x) \cdots p_r(x)$ with irreducibles $p_i(x) \in F[x]$. Then there is $d \in D$ such that $df(x) = q_1(x) \cdots q_r(x)$ with $q_i(x) \in D[x]$. First, $q_i(x)$'s are irreducibles in $F[x]$. Now we write $f(x) = cg(x)$ and $q_i(x) = c_i q_i'(x)$ with $c, c_i \in D$ and primitive $g, q_i' \in D[x]$ for $i = 1, \ldots, r$. Then $dcg(x) = (c_1 \cdots c_r) q_1'(x) \cdots q_r'(x)$. By the uniqueness of a primitive element, $dcu = c_1 \cdots c_r$ for a unit $u \in D$. So, $f(x) = cg(x) = cuq_1'(x) \cdots q_r'(x)$, that is, $f(x)$ factors into irreducibles in $D[x]$. Since the content $c \in D$ and primitive $q_i'(x) \in D[x] \subseteq F[x]$ are unique up to a unit factor, the factorization of $f(x)$ in $D[x]$ is unique. □

EXAMPLE 6.29. $\mathbb{Z}[x]$ and $\mathbb{F}[x, y]$ are UFD's. However, neither is not a PID (e.g. $\langle x, 2\rangle \leq \mathbb{Z}[x]$ and $\langle x, y\rangle \leq \mathbb{F}[x, y]$), even if both $\mathbb{Z}$ and $\mathbb{F}[x]$ are PID's. Note that the multivariate polynomial ring over a field is a complicated object, comparatively with the univariate polynomial ring $\mathbb{F}[x]$.

The last object to be studied in this section is a Euclidean domain. A main example except for $\mathbb{Z}$ and $\mathbb{F}[x]$ is the domain $\mathbb{Z}[i]$ of Gaussian integers.

DEFINITION 6.30. Let D be an integral domain. A function $\nu : D \setminus \{0\} \to \mathbb{Z}_{\geq 0}$ is called a *Euclidean valuation* on D if

i) for each $a, b \in D$ with $b \neq 0$, there exist $q, r \in D$ such that $a = bq + r$ with $r = 0$ or $\nu(r) < \nu(b)$;

ii) for each $a, b \in D \setminus \{0\}$, $\nu(a) \leq \nu(ab)$.

If there is a Euclidean valuation ν on D, we call D a *Euclidean domain*(ED).

EXAMPLE 6.31. $\mathbb{Z}$ with $\nu(n) = |n|$, and $\mathbb{F}[x]$ with $\nu(f) = \deg f$ are ED's. Of course, every division ring(field and skew-field) is a ED, since we take $\nu(a) = 1$ for each nonzero a.

Theorem 6.32. *Every ED is a PID.*

Proof. Let I be an ideal of a ED D. If $I = \{0\}$ then $I = \langle 0 \rangle$. If I is nontrivial, then there is a nonzero $b \in I$ such that $\nu(b) \le \nu(a)$ for each nonzero $a \in I$. We claim that $I = \langle b \rangle$: for each $a \in I$, $a = bq + r$ with $r = 0$ or $\nu(r) < \nu(b)$, so $r = a - bq \in I$ leads us that $r = 0$ because $\nu(r) < \nu(b)$ is not possible from the minimality of $\nu(b)$, thus we proved. □

Corollary 6.33. *Every ED is a UFD.*

EXAMPLE 6.34. * (1) $\mathbb{Z}[i]$, $\mathbb{Z}[\frac{-1+\sqrt{-3}}{2}]$, $\mathbb{Z}[\sqrt{-2}]$ are ED's.

(2) In $\mathbb{Z}[i]$, $52 = (1+i)^2(1-i)^2(3+2i)(3-2i)$ is a factorization into irreducibles.

Proposition 6.35. *Let (D, ν) be a ED. Then $\nu(1) \le \nu(a)$ for each $a \in D \setminus \{0\}$. Also, an element $u \in D$ is a unit if and only if $\nu(u) = \nu(1)$.*

Proof. By the second condition of ν, $\nu(1) \le \nu(1 \cdot a) = \nu(a)$. If $u \in D$ is a unit, then $\nu(u) \le \nu(u \cdot u^{-1}) = \nu(1)$. The converse implication follows from the first condition of ν: $1 = uq + r$ with $r = 0$ or $\nu(r) < \nu(u)$, so $r = 0$, which means that u is a unit. □

DEFINITION 6.36. Let D be a UFD and $a, b \in D$. An element $d \in D$ is a greatest common divisor, denoted by $\gcd(a, b)$, if

i) $d|a$ and $d|b$;
ii) $c|d$ for each $c|a$ and $c|b$.

Proposition 6.37. *Let D be a PID and $a, b \in D \setminus \{0\}$. Then there exists $d = \gcd(a, b)$ and $d = \lambda a + \mu b$ for some $\lambda, \mu \in D$.*

Proof. Let $I = \{ra + sb |\ r, s \in D\}$ be an ideal of D. Then $I = \langle d \rangle$ for some $d \in I$. Since $d | ra + sb$ for each $r, s \in D$, the cases of $(r, s) = (1, 0), (0, 1)$ lead us that $d|a$ and $d|b$. For the next condition, consider any $c \in D$ with $c|a$ and $c|b$. Then we have $c | ra + sb$, i.e., $c|x$ for all $x \in I$. So $c|d$, thus $d = \gcd(a, b)$ and $d = \lambda a + \mu b$ for some $\lambda, \mu \in D$. □

Theorem 6.38 (Euclidean algorithm). *Let (D, ν) be a ED and $a, b \in D \setminus \{0\}$. Assume $\nu(a) \ge \nu(b)$. We divide a by b, and b by the remainder, recursively as* *follow*:

$$a = bq_1 + r_1 \quad \textit{with } r_1 \neq 0,\ \nu(r_1) < \nu(b),$$

$$b = r_1 q_2 + r_2 \quad \textit{with } r_2 \neq 0,\ \nu(r_2) < \nu(r_1),$$

$$\vdots$$

$$r_{i-1} = r_i q_{i+1} + r_{i+1} \quad \textit{with } r_{i+1} \neq 0,\ \nu(r_{i+1}) < \nu(r_i),$$

$$\vdots$$

$$r_{s-1} = r_s q_{s+1} + 0.$$

Then we conclude that $\gcd(a, b) = r_s$.

Proof. Since $\nu(r_{i+1}) < \nu(r_i) \in \mathbb{Z}_{\geq 0}$, we obtain $r_{s+1} = 0$ after finite steps. We can check that the common divisors of a and b are the same as those of b and r_1. Continuing this process, those are identical to the common divisors of r_{s-1} and r_s. So, $\gcd(a, b) = \gcd(b, r_1) = \cdots = \gcd(r_{s-1}, r_s) = r_s$. □

DEFINITION 6.39. We define the set of *Gaussian integers* to be

$$\mathbb{Z}[i] = \{a + bi |\ a, b \in \mathbb{Z}\},$$

and let $N(a + bi) = a^2 + b^2$.

Proposition 6.40. (a) *The norm function* $N : \mathbb{Z}[i] \to \mathbb{Z}_{\geq 0}$ *satisfies that*

i) $N(\alpha) = 0$ *if and only if* $\alpha = 0$, *and*

ii) $N(\alpha\beta) = N(\alpha)N(\beta)$. *We call this norm* N *a* multiplicative norm.

(b) $\mathbb{Z}[i]$ *is an integral domain.*

(c) $\mathbb{Z}[i]$ *is a ED with a valuation* N.

Proof. (b) We show that $\mathbb{Z}[i]$ has no zero divisors. Let $\alpha\beta = 0$. Then $N(\alpha)N(\beta) = N(\alpha\beta) = N(0) = 0$, so $N(\alpha) = 0$ or $N(\beta) = 0$, that is, $\alpha = 0$ or $\beta = 0$.

(c) Let $\alpha = a_1 + a_2 i, \beta = b_1 + b_2 i \in \mathbb{Z}[i]$ with $\beta \neq 0$. Then

$$N(\alpha) \leq N(\alpha)N(\beta) = N(\alpha\beta).$$

The division algorithm holds: we have $\alpha = \beta q + r$ for some $q, r \in \mathbb{Z}[i]$ with $r = 0$ or $N(r) < N(\beta)$, since there exists $q_1 + q_2 i \in \mathbb{Z}[i]$ such that, for $r = (a_1 + a_2 i) - (b_1 + b_2 i)(q_1 + q_2 i)$,

$$N(r) = (a_1 - b_1 q_1 + b_2 q_2)^2 + (a_2 - b_2 q_1 + b_1 q_2)^2 < b_1^2 + b_2^2 = N(\beta).$$

Note that this $q_1 + q_2 i \in \mathbb{Z}[i]$ is unique. □

EXERCISE 6.41. (1) Let D be a ED with a multiplicative norm $N : D \to \mathbb{Z}_{\geq 0}$. Then $a \in D$ is a unit if and only if $N(a) = 1$. If $N(\pi)$ is a prime, then $\pi \in D$ is an irreducible in D.

(2) The units in $\mathbb{Z}[i]$ are $\pm 1, \pm i$. $5 = (1+2i)(1-2i) \in \mathbb{Z}[i]$ is not an irreducible.

(3) Let $D = \mathbb{Z}[\sqrt{-d}] = \{a + b\sqrt{-d} | \ a, b \in \mathbb{Z}\}$ where d is an integer greater than 1. The only units in D are ± 1. [Hint: $N(a+b\sqrt{-d}) = a^2 + db^2$, $N(\alpha\beta) = N(\alpha)N(\beta)$.]

EXERCISE 6.42. (1) Let p be an odd prime in $\mathbb{Z}$. Then $p = a^2 + b^2$ for integers a and b in $\mathbb{Z}$ if and only if $p \equiv 1 \pmod 4$.

(2) 2 is equal to a product of a unit and the square of an irreducible in $\mathbb{Z}[i]$.

(3) An odd prime p in $\mathbb{Z}$ is irreducible in $\mathbb{Z}[i]$ if and only if $p \equiv 3 \pmod 4$.

EXERCISE 6.43. (1) Let $\langle \alpha \rangle$ be a nonzero principal ideal in $\mathbb{Z}[i]$. Then $\mathbb{Z}[i]/\langle \alpha \rangle$ is a finite ring. [Hint: Use the division algorithm.]

(2) If π is an irreducible in $\mathbb{Z}[i]$, then $\mathbb{Z}[i]/\langle \pi \rangle$ is a field. Find the order and characteristic of $\mathbb{Z}[i]/\langle 1+i \rangle$.

EXAMPLE 6.44. * $\mathbb{Z}[\frac{1+\sqrt{-19}}{2}]$ is a PID, but not a ED.[10]

7. Local rings *

DEFINITION 7.1. A commutative ring with 1 having a unique maximal ideal is called a *local ring*.[11]

EXERCISE 7.2. Prove that a nonzero homomorphic image of a local ring is again a local ring.

Theorem 7.3. (a) *R is a local ring if and only if the nonunits in R form an ideal M, which is a unique maximal ideal.*

(b) *Let P be a prime ideal in R and R_P be the localization of R at P. Then*

$$\{\text{prime ideals in } R \text{ contained in } P\} \overset{1-1}{\longleftrightarrow} \{\text{prime ideals in } R_P\}.$$

EXERCISE 7.4. R is a local ring if and only if, for each $r, s \in R$, $r + s = 1$ implies that either r or s is a unit.

---

[10]J. C. Wilson, *A Principal Ring that is Not a Euclidean Ring*, Math. Magazine, 1973, pp.34–38.

[11]The name comes from the fact that these rings are important in the study of the local behavior of varieties and manifolds: the ring of function germs at a point is always local.

EXAMPLE 7.5. (1) The ring $\mathbb{Z}_{\langle p\rangle}$ consisting of all rational numbers with denominators not divisible by a fixed prime p is a local ring.

(2) $\mathbb{F}[x]_{\langle x-a\rangle} = \{\frac{f}{g} | f, g \in \mathbb{F}[x], g(a) \neq 0\}$ is a local ring.

EXERCISE 7.6. (1) Let R be an integral domain and consider R_P as a subring of the quotient field of R. Then $R = \bigcap R_M$, where the intersection is over localizations at each maximal ideal M.

(2) Every nonzero homomorphic image of a local ring is local.

(3) Let M be a maximal ideal of R. Then for a positive integer n, R/M^n has a unique prime ideal M/M^n.

(4) The following are equivalent: (i) R has a unique prime ideal; (ii) Every nonunit in R is nilpotent; (iii) R has a minimal prime ideal which contain all zero divisors, and all nonunits of R are zero divisors.

EXERCISE 7.7. (1) Let $R = \mathbb{F}[x]$ and a multiplicative set $S = \{x^n | \; n = 0, 1, 2, \ldots\}$. Is $S^{-1}R$ a PID? ($S^{-1}R$ is called the ring of Laurent polynomials over $\mathbb{F}$.)

(2) Let R be a UFD and p be a prime element. Then the local ring $R_{(p)}$ is principal.

EXERCISE 7.8. Let I be an ideal of R, $\pi : R \to R/I$ the canonical projection, and S a multiplicative subset of R. Show that $S^{-1}R/S^{-1}I \cong (\pi(S))^{-1}(R/I)$.

EXERCISE 7.9. Let S be a multiplicative subset of R with $0 \notin S$ and $1 \in S$. Describe S^{-1} as a functor and show that it is a exact functor.

CHAPTER 6

Extension fields

In this chapter, we study a basic field theory: field extensions, finite fields. As an application, old questions on geometric constructions are considered.

1. Basics

DEFINITION 1.1. If F and E are fields with $F \subseteq E$, then we say that E is an *extension field* of F, denoted by $F \leq E$.

Let F be a field. Our goal is understanding

how to obtain a field E over which a polynomial $f(x) \in F[x]$ has a zero.

Theorem 1.2 (Kronecker 1881). *Let F be a field and $f(x) \in F[x]$ be a nonconstant polynomial. Then there exists an extension field E of F such that $\alpha \in E$ and $f(\alpha) = 0$.*

Proof. Note that $F[x]$ is a UFD. We let $p(x)$ be an irreducible factor of $f(x)$. The ideal $\langle p(x)\rangle \leq F[x]$ is maximal and $F[x]/\langle p(x)\rangle$ is a field. Consider a map

$$\psi : F \to F[x]/\langle p(x)\rangle \text{ by } a \mapsto a + \langle p(x)\rangle,$$

which is a monomorphism. Set $\alpha = x + \langle p(x)\rangle \in F[x]/\langle p(x)\rangle =: E$. Then $p(\alpha) = 0 \in E$. □

EXAMPLE 1.3. (1) The polynomial $p(x) = x^2 + 1 \in \mathbb{R}[x]$ is irreducible, so that $\mathbb{R}[x]/\langle x^2+1\rangle$ is a field. By letting $\alpha = x+\langle x^2+1\rangle$, $p(\alpha) = \alpha^2+1 = 0 \in \mathbb{R}[x]/\langle x^2+1\rangle$. Then

$$\mathbb{R}[x]/\langle x^2 + 1\rangle = \{a + b\alpha |\ \alpha^2 + 1 = 0,\ a, b \in \mathbb{R}\} =: \mathbb{R}(\alpha) \cong \mathbb{C}.$$

(2) How to obtain E such that $x^2 - 2 \in \mathbb{Q}[x]$ has a zero in E?

DEFINITION 1.4. (a) Let E be an extension field of F. An element $\alpha \in E$ is called *algebraic over F* if there is a nonconstant polynomial $f(x) \in F[x]$ such that $f(\alpha) = 0$. We say that $\alpha \in E$ is *transcendental over F* if it is not algebraic.

(b) A complex number α is called an *algebraic number* if it is algebraic over $\mathbb{Q}$. If it is not, $\alpha \in \mathbb{C}$ is a *transcendental number.*

EXAMPLE 1.5. The number π is the ratio of the circumference of a circle to its diameter. We know that neither e nor π is algebraic.[1] In fact, it is known by Georg Cantor[2] that the transcendental numbers are uncountably many.

Theorem 1.6. *Let $F \leq E$ be extension fields and $\alpha \in E$. Consider the evaluation homomorphism $\phi_\alpha : F[x] \to E$. Then α is transcendental over F if and only if ϕ_α is 1-1.*

Proof. For any nonconstant polynomial $f(x) \in F[x]$, $f(\alpha) \neq 0$, i.e., $\phi_\alpha(f(x)) \neq 0$ if and only if $\ker \phi_\alpha = \{0\}$. □

Theorem 1.7. *Let $F \leq E$ be extension fields and $\alpha \in E$ be algebraic over F. Then there is an irreducible polynomial $p(x) \in F[x]$ such that $p(\alpha) = 0$.*

This $p(x)$ of minimal degree having α as a zero and is unique up to a constant factor in F. We denote the monic(leading coefficient=1) $p(x)$ by $\mathrm{irr}(\alpha, F)$ and the degree of $p(x)$ by $\deg(\alpha, F)$.

Proof. Let $\phi_\alpha : F[x] \to E$ be the evaluation homomorphism at α. Then $\ker \phi_\alpha$ is an ideal in a principal ideal domain $F[x]$. So $\ker \phi_\alpha = \langle p(x) \rangle$ for some $p(x) \in F[x]$. For any nonconstant polynomial $f(x) \in F[x]$ with $f(\alpha) = 0$, $f(x) \in \langle p(x) \rangle$, that is, $p(x) | f(x)$. Thus, $p(x)$ is of minimal degree having α as a zero. If $\deg f(x) = \deg p(x)$ then $f(x) = ap(x)$ for $a \in F$. Suppose that $p(x)$ factors into $p(x) = r(x)s(x)$ of lower degrees, then $0 = p(\alpha) = r(\alpha)s(\alpha)$ leads us that $r(\alpha) = 0$ or $s(\alpha) = 0$, which contradicts to the minimality of $\deg p(x)$ with $p(\alpha) = 0$. Hence, $p(x) \in F[x]$ is irreducible. □

EXAMPLE 1.8. (1) $\mathrm{irr}(\sqrt{2}, \mathbb{Q}) = x^2 - 2$, but $\mathrm{irr}(\sqrt{2}, \mathbb{R}) = x - \sqrt{2}$.

(2) $\mathrm{irr}(\sqrt{1+\sqrt{3}}, \mathbb{Q}) = x^4 - 2x^2 - 2$.

---

[1] By Hermite(1822–1901, a French mathematician) in 1873, and by Lindemann(1852–1939, a German mathematician) in 1882, respectively. However, it has not been proved yet whether $e + \pi$, $e \cdot \pi$, π^e are transcendental, although we expect they are.

[2] Georg Ferdinand Ludwig Philipp Cantor(1845–1918), a German mathematician, the inventer of set theory. He discovered a hierarchy in transfinite numbers.

For extension fields $F \le E$ and $\alpha \in E$, we define

$$F(\alpha), \text{ the smallest subfield of } E \text{ containing } F \text{ and } \alpha.$$

Let $\phi_\alpha : F[x] \to E$ be the evaluation homomorphism at α.

(Case 1: α is algebraic over F) Since $\ker \phi_\alpha = \langle \mathrm{irr}(\alpha, F) \rangle$ is a maximal ideal of $F[x]$, $F[x]/\langle \mathrm{irr}(\alpha, F)\rangle$ is a field, which is isomorphic to $\phi_\alpha(F[x]) \le E$. We define

$$F[x]/\langle \mathrm{irr}(\alpha, F)\rangle \cong \phi_\alpha(F[x]) =: F(\alpha).$$

(Case 2: α is transcendental over F) Since ϕ_α is 1-1, $F[x]$ is isomorphic to $\phi_\alpha(F[x]) \le E$. The integral domain $\phi_\alpha(F[x])$ is denoted by $F[\alpha]$. The field of quotients of $F[\alpha]$ is a subfield of E, which we denote by $F(\alpha)$. Note that $F(\alpha)$ is isomorphic to the quotient field $F(x)$ of $F[x]$.

Exercise 1.9. (1) Show that if $\alpha \in E$ is algebraic over F then there is an isomorphism of fields $\phi : F(\alpha) \to F[x]/\langle \mathrm{irr}(\alpha, F)\rangle$, where $F(\alpha)$ is the smallest subfield of E containing F and α.

(2) Show that if $\alpha \in \mathbb{C}$ is transcendental over $\mathbb{Q}$ then $\mathbb{Q}[\alpha]$ is not a field.

(3) Show that if $\alpha \in E$ is transcendental over F then there is an isomorphism of fields $\phi : F(\alpha) \to F(x)$ such that $\phi(\alpha) = x, \phi|_F = \mathrm{id}$.

Exercise 1.10. Let E be an extension field of F and $u \in E$ be transcendental over F. Let K be a subfield of $F(u)$ properly containing F and choose an element $v \in F(u) \setminus F$. Show that v is transcendental over F and that u is algebraic over K.

Solution. Since $u \in F(u) \setminus F$ is transcendental over F, so is u^m $(m \in \mathbb{Z} \setminus \{0\})$, and generally so is $f(u)/g(u)$, a rational function in $F(u) \setminus F$ ($\because$ Suppose that there exists a nonzero polynomial such that $f(u)/g(u)$ is a root. Then by multiplying the leading denominator, there exists a nonzero polynomial with u as a zero.). Hence any $v \in F(u) \setminus F$ is transcendental over F.

Since K contains an element v_0 which is transcendental over F, we have

$$F \lneq F(v_0) \le K \le F(u)$$

where v_0 is a rational function $f(u)/g(u)$. So we get $\mathrm{irr}(u, F(v_0))$ and

$$[F(u) : F(v_0)] = \deg(u, F(v_0)) < \infty.$$

Hence $[F(u) : K] < \infty$ and u is algebraic over K. $\square$

DEFINITION 1.11. A field $E \geq F$ is a simple extension of F if $E = F(\alpha)$ for some $\alpha \in E$.

Theorem 1.12. *Let a field $E = F(\alpha)$ is a simple extension of F with $\alpha \in E$ algebraic over F, and let $n = \deg(\alpha, F) \geq 1$. Then, for each $\beta \in E = F(\alpha)$, there exist unique $b_0, \ldots, b_{n-1} \in F$ such that*

$$\beta = b_0 + b_1\alpha + \cdots + b_{n-1}\alpha^{n-1}.$$

REMARK 1.13. Let α be algebraic over a field F. Then $F(\alpha)$ is a vector space over F with basis $\{1, \alpha, \alpha^2, \ldots, \alpha^{n-1}\}$, where $n = \deg(\alpha, F)$.

EXAMPLE 1.14. (1) Let $\mathbb{Q} < \mathbb{Q}(\sqrt[4]{2}) < \mathbb{R}$. Then $\mathrm{irr}(\sqrt[4]{2}, \mathbb{Q}) = x^4 - 2$ and the field $\mathbb{Q}(\sqrt[4]{2})$ is a vector space with basis $\{1, \sqrt[4]{2}, \sqrt{2}, (\sqrt[4]{2})^3\}$.

(2) Let $E = \mathbb{Z}_2[x]/\langle x^2 + x + 1\rangle \hookleftarrow \mathbb{Z}_2$. Note that $x^2 + x + 1$ is irreducible over $\mathbb{Z}_2$, so $E \cong \mathbb{Z}_2(\alpha)$ is a field with $|E| = 2^2 = 4$, where α corresponds to $x + \langle x^2 + x + 1\rangle$. Find the multiplication table of the finite field of order 2^2.

Theorem 1.15. *Let $F \leq E$ be extension fields and $\alpha \in E$ be algebraic over F with $\deg(\alpha, F) = n$. Then $F(\alpha)$ has a basis $\{1, \alpha, \alpha^2, \ldots, \alpha^{n-1}\}$ over F, and each $\beta \in F(\alpha)$ is algebraic over F with $\deg(\beta, F) \leq \deg(\alpha, F)$.*

EXERCISE 1.16. * Let $F = \mathbb{Q}(\sqrt{2})$. Consider an action of F on itself defined by multiplication. For each $\alpha \in F$, one may view the action of α as a linear operator ϕ_α on F, where F is regarded as a vector space over $\mathbb{Q}$. Describe the trace and determinant of ϕ_α. [3]

EXERCISE 1.17. * Let F/k be extension fields and let $k[x]$ be the polynomial ring over k.

(1) Is $F \otimes_k k[x]$ isomorphic to $F[x]$ as F-modules?

(2) Is $F \otimes_k k[x]$ isomorphic to $F[x]$ as F-algebras?

EXERCISE 1.18. * (1) Show that $\mathbb{Z}[x] \otimes_{\mathbb{Z}} \mathbb{Z}_m \cong \mathbb{Z}_m[x]$ as commutative rings.

(2) Show that $\mathbb{Z}[x]/\langle x^3 + x^2 + 1\rangle \otimes_{\mathbb{Z}} \mathbb{Z}_5$ is isomorphic to an extension field of $\mathbb{Z}_5$.

---

[3] Answer: $\mathrm{tr}\,\phi_\alpha = \alpha + \overline{\alpha}$, $\det \phi_\alpha = \alpha\overline{\alpha}$.

2. Algebraic extensions

DEFINITION 2.1. (a) A field $E \geq F$ is called an *algebraic* extension of F if each element of E is algebraic over F.

(b) We say that a field $E \geq F$ is a *finite* extension of degree n over F if $\dim E = n < \infty$ as a vector space over F. The degree of E over F is denote by $[E, F]$.

Theorem 2.2. *Every finite extension of fields is algebraic.*

Proof. Let $[E : F] = n$ and $u \in E$ be arbitrary. Then the set of $n+1$ elements $\{1, u, u^2, \ldots, u^n\}$ is linearly dependent over F. That is, there exist $a_i \in F$, not all zero, such that $a_0 + a_1 u + a_2 u^2 + \cdots + a_n u^n = 0$, which means that u is algebraic over F. □

EXAMPLE 2.3. $\mathbb{Q} < \mathbb{Q}(\sqrt{2}, 2^{\frac{1}{3}}, 2^{\frac{1}{4}}, \ldots)$ is an infinite algebraic extension. Note that

$$\mathbb{Q} < \mathbb{Q}(\sqrt{2}, 2^{\frac{1}{3}}, 2^{\frac{1}{4}}, \ldots) < \overline{\mathbb{Q}}_{\mathbb{C}} < \mathbb{C},$$

where $\overline{\mathbb{Q}}_{\mathbb{C}}$ is the field of algebraic numbers.

Theorem 2.4. *Let $F \leq E$ and $E \leq K$ be finite extensions of fields. Then $F \leq K$ is a finite extension and $[K : F] = [K : E][E : F]$.*

Proof. Let $\{\alpha_1, \ldots, \alpha_n\}$ be a basis of E over F, and $\{\beta_1, \ldots, \beta_m\}$ be a basis of K over E. Each $\gamma \in K$ is uniquely written as $\gamma = \sum_{j=1}^{m} b_j \beta_j$ with $b_j \in E$, and the elements b_j's are uniquely written as $b_j = \sum_{i=1}^{n} a_{ij} \alpha_i$ for $a_{ij} \in F$. Thus we combine two expressions to obtain that

$$\gamma = \sum_{j=1}^{m} \left(\sum_{i=1}^{n} a_{ij} \alpha_i \right) \beta_j = \sum_{i,j} a_{ij} \alpha_i \beta_j$$

for unique $a_{ij} \in F$. That is, we showed that $\{\alpha_i \beta_j | \ 1 \leq i \leq n, 1 \leq j \leq m\}$ is a basis of K over F. □

Corollary 2.5. (a) *Let $F_i \leq F_{i+1}$ be finite extensions of fields for $i = 1, \ldots, r-1$. Then $F_1 \leq F_r$ is a finite extension and $[F_r : F_1] = \prod_{i=1}^{r-1} [F_{i+1} : F_i]$.*

(b) *Let $\alpha \in E \geq F$ be algebraic over F, and $\beta \in F(\alpha) \leq E$. Then* $\deg(\beta, F) | \deg(\alpha, F)$.

DEFINITION 2.6. Let $\alpha_i \in E \geq F$ for $i = 1, \ldots, n$. The field $F(\alpha_1, \ldots, \alpha_n)$ is the smallest subfield of E containing F and α_i's.

EXAMPLE 2.7. (1) A basis of $\mathbb{Q}(\sqrt{2})$ over $\mathbb{Q}$ is $\{1, \sqrt{2}\}$.

(2) Compare $\mathbb{Q}(\sqrt{2}+\sqrt{3})$ with $\mathbb{Q}(\sqrt{2}, \sqrt{3})$. First, we know that

$$[\mathbb{Q}(\sqrt{2}+\sqrt{3}) : \mathbb{Q}] = 4 \text{ and } \mathbb{Q}(\sqrt{2}+\sqrt{3}) \leq \mathbb{Q}(\sqrt{2}, \sqrt{3}).$$

Since $\sqrt{3} \notin \mathbb{Q}(\sqrt{2})$, $\{1, \sqrt{3}\}$ is a basis of $\mathbb{Q}(\sqrt{2}, \sqrt{3})$ over $\mathbb{Q}(\sqrt{2})$. So $\{1, \sqrt{2}, \sqrt{3}, \sqrt{6}\}$ is a basis of $\mathbb{Q}(\sqrt{2}, \sqrt{3})$ over $\mathbb{Q}$, leading us that

$$[\mathbb{Q}(\sqrt{2}, \sqrt{3}) : \mathbb{Q}] = 4 \text{ and } \mathbb{Q}(\sqrt{2}+\sqrt{3}) = \mathbb{Q}(\sqrt{2}, \sqrt{3}).$$

EXERCISE 2.8. (1) Show that $\mathbb{Q}(\sqrt[3]{2}+i) = \mathbb{Q}(\sqrt[3]{2}, i)$.

(2) Show that $\mathbb{Q}(\sqrt[4]{2}) = \mathbb{Q}(\sqrt{2}+\sqrt[4]{2})$.

Theorem 2.9. *Let $F \leq E$ be an algebraic extension. Then there exist finitely many $\alpha_i \in E$ such that $F(\alpha_1, \ldots, \alpha_n) = E$ if and only if $[E : F] < \infty$.*

Proof. $\Rightarrow$) It follows from a tower of algebraic extensions

$$F \leq F(\alpha_1) \leq F(\alpha_1, \alpha_2) \leq \cdots \leq F(\alpha_1, \ldots, \alpha_n).$$

$\Leftarrow$) If $[E : F] = 1$ then $F(1) = E$. If $F \lneq E$ then take $\alpha_1 \in E \setminus F$, that is, $[F(\alpha_1) : F] > 1$. If $F(\alpha_1) \lneq E$ then take $\alpha_2 \in E \setminus F(\alpha_1)$. Since $[E : F] < \infty$, this process stops in a finite number of steps. □

EXERCISE 2.10. (1) Prove that an integral domain D which is a finite-dimensional vector space over a field F is a field.
[Hint: Since $[F[u] : F] < \infty$ for $u \in D$, there exists $\operatorname{irr}(u, F)$.]

(2) Let u be the real root of $x^3 - 6x^2 - 9x + 3$. Write $\alpha = 3u^5 - u^4 + 2$ as a linear combination of a basis of $\mathbb{Q}(u)$ over $\mathbb{Q}$. Find the inverse of $\beta = u^2 - 6u + 8 \in \mathbb{Q}(u)$.

(3) Let E be an extension field of F and $u, v \in E$. Let $f(x)$ be an irreducible polynomial in $F[x]$. Show that u, v are roots of $f(x)$ if and only if there is an isomorphism of fields $\phi : F(u) \cong F(v)$ such that $\phi(u) = v, \phi|_K = \text{id}$.

Theorem 2.11. *$F \leq E$ and $E \leq K$ are algebraic extensions of fields if and only if K is an algebraic extension of F.*

Proof. The proof of if-part is trivial.

$\Rightarrow$) Let $\alpha \in K$. Since α is algebraic over E, we have

$$\operatorname{irr}(\alpha, E) = a_0 + a_1 x + \cdots + a_{n-1} x^{n-1} + x^n \in E[x].$$

Note that each $a_i \in E$ $(i = 0, \ldots, n-1)$ is algebraic over F, that is, $F(a_0, \ldots, a_{n-1})$ is a finite extension of F of some degree $m < \infty$. From the fact that

$$\begin{aligned}[F(a_0, \ldots, a_{n-1}, \alpha) : F] &= [F(a_0, \ldots, a_{n-1}, \alpha) : F(a_0, \ldots, a_{n-1})][F(a_0, \ldots, a_{n-1}) : F] \\ &= nm < \infty,\end{aligned}$$

$\alpha \in F(a_0, \ldots, a_{n-1}, \alpha)$ is a finite extension of F, so is algebraic over F. □

EXERCISE 2.12. (1) Let E be an extension field of F. If $[E : F]$ is a prime, then show that there are no field properly between F and E and E is a simple extension of F, indeed, $E = F(u)$ for every $u \in E \setminus F$.

(2) Let E be an algebraic extension field of F. Let D be an integral domain such that $F \subseteq D \subseteq E$. Show that D is a field.

(3) Let $\alpha = \sqrt[3]{2} + \sqrt{3}$. Find the irreducible polynomial of α over $\mathbb{Q}$ and prove that $\mathbb{Q}(\alpha) = \mathbb{Q}(\sqrt[3]{2})\mathbb{Q}(\sqrt{3})$.

(4) Let $F = \mathbb{Q}(i, \sqrt{3}, \omega)$, where ω is a primitive 3rd root of 1. Find $[F : \mathbb{Q}]$ and a basis of F over $\mathbb{Q}$.

EXERCISE 2.13. Let E be an extension field of F. Let $u \in E$ be algebraic of odd degree over F. Show that u^2 is algebraic of odd degree over F and $F(u) = F(u^2)$.

Solution. Clearly, $F(u^2) \leq F(u)$. Suppose $F(u^2) \lneq F(u)$. Then, since u is a zero of $x^2 - u^2 \in F(u^2)[x]$, we get $[F(u) : F(u^2)] = 2$. But, 2 cannot divide $[F(u) : F]$ ($\because$ $[F(u) : F]$ is odd), which is a contradiction. So we have $F(u^2) = F(u)$ and u^2 is also algebraic over F of degree equal to $\deg(u, F)$. □

EXERCISE 2.14. Show that E is an algebraic extension of F if and only if for every intermediate field K every monomorphism $\phi : K \to K$ with $\phi|_F = \mathrm{id}$ is actually an automorphism of K. [Hint: ($\Rightarrow$) Suppose there exists $u \in K \setminus \mathrm{Im}\,\phi$. Then $u, \phi(u), \phi^2(u), \ldots$ are zeros of $\mathrm{irr}(u, F)$. ($\Leftarrow$) Suppose that E/F is transcendental. Let $K = F(u)$ where u is transcendental, and $\phi : K \to K$ with $\phi(u) = u^2$.]

Theorem 2.15. *Let E be an extension field of F and $K = \{u \in E|\ u$ is algebraic over $F\}$. Then K is a subfield of E. This K is called the* algebraic closure *of F in E, denote by $\overline{F_E}$.*

Proof. Choose any $u, v \in K$ $(v \neq 0)$. Since $F(u, v)$ is finite-dimensional over F by considering $F \leq F(u) \leq F(u, v)$, $F(u, v)$ is an algebraic extension of F.

So we get $F(u,v) \subseteq K$ from the definition. Hence $u - v \in F(u,v) \subseteq K$ and $uv^{-1} \in F(u,v) \subseteq K$, which implies that K is a field. $\square$

DEFINITION 2.16. A field F is called *algebraically closed* if every nonconstant $f(x) \in F[x]$ has a zero in F.

Theorem 2.17. *A field F is algebraically closed if and only if every nonconstant $f(x) \in F[x]$ factors in $F[x]$ into linear factors.*

Proof. $\Rightarrow$) Let $f(x)$ be a nonconstant polynomial in $F[x]$. Since $f(x)$ has zero $a \in F$, $f(x) = (x-a)g(x)$ with $g(x) \in F[x]$. If $g(x)$ is not constant, then $g(x) = (x-b)h(x)$ with $b \in F$ and $h(x) \in F[x]$. Continuing this process, we get a factorization into linear polynomials.

$\Leftarrow$) If $f(x)$ has a linear factor $ax - b \in F[x]$, then $a^{-1}b \in F$ is a zero of $f(x)$. $\square$

Corollary 2.18. *A field F is algebraically closed if and only if F has no proper algebraic extension.*

Proof. $\Rightarrow$) Let $F \leq E$ be an algebraic extension. For each $\alpha \in E$, we have $\operatorname{irr}(\alpha, F) = x - \alpha \in F[x]$ since F is algebraically closed. Thus $\alpha \in F$.

$\Leftarrow$) Suppose that there is $f(x) \in F[x]$ with $\deg f(x) \geq 2$ such that $f(x)$ has no zero in F. Then $f(x) = f_1(x) \cdots f_k(x)$ where $f_i(x)$'s are irreducible of degree greater than 1. Consider $E := F[x]/\langle f_1(x)\rangle \cong F(\alpha)$ with $\alpha = x + \langle f_1(x)\rangle$. Then $[F(\alpha) : F] = \deg f_1(x) \geq 2$, which is a contradiction. $\square$

Theorem 2.19. *

(a) *Every field F has an algebraic closure $\overline{F}$, an algebraic extension of F which is algebraically closed.*

(b) *The field $\mathbb{C}$ of complex numbers is algebraically closed.*

EXERCISE 2.20. * Let $\overline{\mathbb{Q}}$ be the subset of $\mathbb{C}$ consisting of all algebraic elements over $\mathbb{Q}$. Prove that $\overline{\mathbb{Q}}$ is an algebraic closure of $\mathbb{Q}$. Is $\overline{\mathbb{Q}}$ a unique algebraic closure of $\mathbb{Q}$ in $\mathbb{C}$? Justify your answer.

Theorem 2.21. (Frobenius)* [4] *Let D be a division ring over $\mathbb{R}$. Then D is isomorphic to $\mathbb{R}, \mathbb{C}$ or $\mathbb{H}$.*

---

[4] For proof, see [Hr, pp.369–371].

3. Finite fields

A finite field is an object which cannot be too emphasized, in particular, with the strongly growing power of computers. In this section, we study the structure of finite fields.

Theorem 3.1. (a) *Let F be a finite field of order q and E be an extension field of F with $[E:F]=n<\infty$. Then $|E|=q^n$.*

(b) *Let $\mathbb{F}_p$ be the field of integers mod p (p, a prime). If $f(x)\in\mathbb{F}_p[x]$ is irreducible over $\mathbb{F}_p$ of degree n, then $\mathbb{F}_p[x]/\langle f(x)\rangle$ is a field with p^n elements.*

Proof. (a) Let $\{\alpha_1,\ldots,\alpha_n\}$ be a basis of E as a vector space over F. Then each $\beta\in E$ is written uniquely as $\beta=b_1\alpha_1+\cdots+b_n\alpha_n$ for $b_i\in F$. That is, E is in 1-1 correspondence to F^n.

(b) $\mathbb{F}_p[x]/\langle f(x)\rangle\cong E$ with $[E:\mathbb{F}_p]=n$. □

Corollary 3.2. *The finite field E with $\operatorname{ch}E=p$ must be of order p^n for some $n\in\mathbb{Z}_+$.*

Proof. By Proposition 3.12 in Ch.5, there is a subfield $F\cong\mathbb{F}_p$. □

EXAMPLE 3.3. Consider $\mathbb{F}_2[x]/\langle f(x)\rangle$, where $f(x)$ is the irreducible polynomial

$$f(x)=x^2+x+1\in\mathbb{F}_2[x].$$

Note that its elements are

$$0,1,\alpha,1+\alpha$$

where $\alpha=x+\langle x^2+x+1\rangle$. The multiplication table is:

$\times$	0	1	α	$1+\alpha$
0	0	0	0	0
1	0	1	α	$1+\alpha$
α	0	α	$1+\alpha$	1
$1+\alpha$	0	$1+\alpha$	1	α

EXERCISE 3.4. (1) Find an irreducible polynomial of degree 3 in $\mathbb{F}_2[x]$. Using this polynomial, find a field of 8 elements. What is the multiplication table?

(2) Construct a field with 16 elements.

(3) Find a field of 9 elements and of 27 elements.

We prove that, for any prime p and $n\in\mathbb{Z}_+$, there exists a finite field of order p^n.

Lemma 3.5. *Let $|F| < \infty$ with* $\mathrm{ch}\, F = p$. *Then $x^{p^n} - x$ has p^n distinct zeros in $\overline{F}$.*

Proof. All zeros of $f(x) = x^{p^n} - x$ are of multiplicity 1, since $f'(x) = -1 \neq 0$ for all $x \in \overline{F}$. □

Theorem 3.6. *Let p be a prime and $n \geq 1$ an integer. Then F is a finite field with p^n elements if and only if*

$$F \cong \{x \in \overline{\mathbb{F}_p} |\ x \text{ is a zero of } x^{p^n} - x \in \mathbb{F}_p[x]\},$$

that is, F is isomorphic to a splitting field of $x^{p^n} - x$ over $\mathbb{F}_p$.

Proof. We have only to show the only-if part. The multiplicative group $F^\times = F \setminus \{0\}$ is of order $p^n - 1$. For each element $\alpha \in F^\times$, we have $|\alpha| | p^n - 1$, so that $\alpha^{p^n - 1} = 1$. Thus, $\alpha^{p^n} = \alpha$ for $\alpha \in F$. Since $x^{p^n} - x$ has p^n zeros, we proved. □

By Theorem 3.6, the field of every prime power order exists.

Definition 3.7. The field F with $|F| = p^n$ (p, a prime) is called the *Galois field* of order p^n, denoted by $GF(p^n)$ or simply $\mathbb{F}_{p^n}$.

Definition 3.8. An element α is an nth root of unity if $\alpha^n = 1$. We say that α is a primitive nth root of unity if $\alpha^n = 1$ and $\alpha^m \neq 1$ for $0 < m < n$.

Theorem 3.9. *Let $|F| < \infty$ and $[E : F] < \infty$. Then $E = F(\alpha)$ for some $\alpha \in E$.*

Proof. By Corollary 5.13 in Ch.4, $E^\times$ is a cyclic multiplicative group. For a generator $\alpha \in E^\times$, $F(\alpha) = E$. □

Example 3.10. (1) $\mathbb{F}_{11}^\times$ is a cyclic multiplicative group of order 10. The generators of $\mathbb{F}_{11}^\times$ are 10th primitive roots of unity in $\mathbb{F}_{11}$:

$$\{2, 8, 7, 6\} = \{2^1, 2^3, 2^5, 2^7\}.$$

The primitive 5th roots of unity are: $\{2^2, 2^4, 2^6, 2^8\} = \{4, 5, 9, 3\}$. The primitive square(2nd) root of unity is $2^5 = 10 = -1 \in \mathbb{F}_{11}$.

(2) Consider $\mathbb{F}_{121} = GF(11^2)$, which is the extension field of $\mathbb{F}_{11}$ of degree 2. Take $\alpha \in \mathbb{F}_{121} \setminus \mathbb{F}_{11}$. Then $\mathbb{F}_{121} = \mathbb{F}_{11}(\alpha)$.

Exercise 3.11. Let F be a finite field. Show that $F = \mathbb{F}_p(u)$ for some element u in F and prime integer p.

Corollary 3.12. *Let F be a finite field and $n \in \mathbb{Z}_+$. Then there exists an irreducible polynomial over F of degree n.*

Proof. Let $|F| = p^r$ where $p = \operatorname{ch} F$. There is a field

$$E = \{x \in \overline{\mathbb{F}_p} |\ x \text{ is a zero of } x^{p^{rn}} - x\} \geq F.$$

Since $[E : F] = n$ and $E = F(\alpha)$ for some $\alpha \in E$, the polynomial $\operatorname{irr}(\alpha, F)$ is irreducible of degree n. □

EXAMPLE 3.13. (1) $x^2 + 1$ is irreducible over $\mathbb{F}_p$ for any prime $p \equiv 3 \pmod 4$. So,

$$\mathbb{F}_{11}[x]/\langle x^2 + 1\rangle$$

is a unique field of order $11^2 = 121$, up to isomorphism.

(2) $x^2 + x + 1$ is irreducible over $\mathbb{F}_p$ for any prime $p \equiv 2 \pmod 3$. So,

$$\mathbb{F}_{29}[x]/\langle x^2 + x + 1\rangle$$

is a unique field of order $29^2 = 841$, up to isomorphism.

Proof. (1) Assume $p > 2$. We know that $x^2 + 1$ has a linear factor $x - \alpha$ if and only if $\alpha^2 + 1 = 0$, that is, $\alpha \neq \pm 1$ and $\alpha^4 = 1$. Since $\mathbb{F}_p^\times$ is a cyclic group of order $p - 1$, there is $\alpha \in \mathbb{F}_p^\times$ whose order is 4 if and only if $p - 1 \equiv 0 \pmod 4$.

Thus, $x^2 + 1$ is irreducible over $\mathbb{F}_p$ if and only if $p - 1 \equiv 2 \pmod 4$.

(2) Assume $p > 3$. Note that $x^2 + x + 1$ has a linear factor $x - \alpha$ if and only if $\alpha^2 + \alpha + 1 = 0$, that is, $\alpha^3 - 1 = (\alpha - 1)(\alpha^2 + \alpha + 1) = 0$. So there is $\alpha \in \mathbb{F}_p^\times$ whose order is 3 if and only if $p - 1 \equiv 0 \pmod 3$.

Hence $x^2 + 1$ is irreducible over $\mathbb{F}_p$ if and only if $p - 1 \equiv 1 \pmod 3$. □

EXERCISE 3.14. Let F be a finite field of prime characteristic p and $f_p : F \to F$ be the *Frobenius map* (i.e., $f_p(x) = x^p$). Then f_p is an isomorphism, and every polynomial of the form $x^p - a \in F[x]$ is not irreducible for any $a \in F$.

Let $q = p^\ell$.

How many irreducible polynomials $f(x) \in \mathbb{F}_q[x]$ of degree n?

This question is connected with the number of generators in $\mathbb{F}_{q^n}$ over $\mathbb{F}_q$. Precisely, it is the number of conjugacy classes of $\alpha \in \overline{\mathbb{F}_q}$ such that $[\mathbb{F}_q(\alpha) : \mathbb{F}_q] = n$. Recall the Frobenius automorphism f_p by $f_p(x) = x^p$, and let $f_q = f_p^\ell$ in $\mathbb{F}_{q^n}$.

Proposition 3.15. *The collection of monic irreducible polynomials $f(x) \in \mathbb{F}_q[x]$ of degree n is in 1-1 correspondence to the collection of sets of n mutually conjugate generators of $\mathbb{F}_{q^n}$ over $\mathbb{F}_q$:*

$$(x-\alpha)(x-\alpha^q)\cdots(x-\alpha^{q^{n-1}}) \overset{1-1}{\longleftrightarrow} \{\alpha, \alpha^q, \dots \alpha^{q^{n-1}}\}.$$

Define the *Möbius*[5] *function* $\mu : \mathbb{Z}_+ \to \mathbb{Z}_+$ by

$$\mu(n) = \begin{cases} (-1)^t & \text{if distinct } t \text{ primes divide } n \text{ but no square does,} \\ 0 & \text{if the square of any prime divides } n. \end{cases}$$

Corollary 3.16. *The number of irreducible polynomials over $\mathbb{F}_q$ of degree n is*

$$\frac{1}{n}\left(\sum_{d|n} \mu(d) q^{n/d}\right).$$

Proof. We remove from $\mathbb{F}_{q^n}$ elements which generate proper subfields of $\mathbb{F}_{q^n}$ over $\mathbb{F}_q$, and divide by n, the number of conjugates of a given generator, so that we obtain the number of sets of n mutually conjugate generators.

Note that there is a subfield $\mathbb{F}_{q^m} \le \mathbb{F}_{q^n}$ if and only if $m|n$. So the maximal subfields of $\mathbb{F}_{q^n}$ are $\mathbb{F}_{q^{n/r}}$ with prime r dividing n.

If $n = r_1 r_2$, the product of distinct primes, then the number of generators is

$$q^n - q^{n/r_1} - q^{n/r_2} + q^{n/(r_1 r_2)}.$$

In a general n, it is exactly $\sum_{d|n} \mu(d) q^{n/d}$. □

Example 3.17. (1) For the case of $q = 2$ and $n = 8$, the number of irreducibles is $\frac{1}{8}\left(2^8 - 2^4\right) = 30$.

(2) What about $q = 2$ and $n = 10$? $\frac{1}{10}\left(2^{10} - 2^5 - 2^2 + 2\right) = 99$.

Example 3.18. Consider the finite field $\mathbb{F}_{256}$ of $256 = 2^8$ elements. As a ring,

$$\mathbb{F}_{256} \cong \mathbb{F}_2[x]/\langle x^8 + x^4 + x^3 + x^2 + 1\rangle.$$

One can use another irreducible polynomial of degree 8, but this is the standard representation of $\mathbb{F}_{256}$ in computer science to represent bytes, namely

$$(a_7, \dots, a_0) = \sum_{n=0}^{7} a_n x^n \in \mathbb{F}_{256}$$

where a_i's are 0 or 1.

[5]August Ferdinand Möbius(1790–1868), a German mathematician and theoretical astronomer.

Final Exam

(June 19, 2014)

1. (30 points) Determine whether each statement is true or not: True(T) or False(F). (If you answer correctly, you get 3 points; If nothing, 1 point; If wrong, 0 point.)
   1) Every subring of a ring R is an ideal of R.
   2) Every prime ideal of a ring is a maximal ideal.
   3) For a commutative ring R with 1, an ideal M is a maximal ideal of R if and only if R/M is a field.
   4) Every ideal in the polynomial ring $\mathbb{F}[x]$ over a field $\mathbb{F}$ is principal.
   5) Let $p(x) \in \mathbb{F}[x]$. Then $\langle p(x) \rangle$ is a maximal ideal if and only if $p(x)$ is irreducible over $\mathbb{F}$.
   6) Every real number is algebraic over $\mathbb{Q}$.
   7) Let $F \leq E$ and $E \leq K$ be finite extensions of fields. Then $F \leq K$ is a finite extension and $[K : F] = [K : E][E : F]$.
   8) Let $\alpha \in E \geq F$ be algebraic over a field F, and $\beta \in F(\alpha) \leq E$. Then $\deg(\beta, F) | \deg(\alpha, F)$.
   9) For each prime p and a positive integer n, there is a unique field of order p^n up to isomorphism.
   10) Let F be a finite field and $n \in \mathbb{Z}_+$. Then there exists an irreducible polynomial over F of degree n.

2. (15pts) (1) Write down the definition of a prime ideal of a ring.
   (2) Find all prime ideals of $\mathbb{Z}$, where $\mathbb{Z}$ is the integral domain of integers.
   (3) Find all maximal ideals of $\mathbb{C}[x]$, where $\mathbb{C}$ is the field of complex numbers.
3. (25pts) (1) Write all prime fields.
   (2) State the Kronecker theorem.
   (3) Construct the finite field $\mathbb{F}_4$ of order 4 and write its multiplication table.

- For each of the following problems, you should justify your answer.

4. (20pts) (1) Find the kernel of a ring homomorphism $\phi : \mathbb{R}[x] \to \mathbb{C}$ defined by $\phi(1) = 1$ and $\phi(x) = i$.
   (2) Using the fundamental homomorphism theorem in (1), which ring is isomorphic to $\mathbb{C}$?
5. (10pts) Show that $\mathbb{Q}(\sqrt{2} + \sqrt{3}) = \mathbb{Q}(\sqrt{2}, \sqrt{3})$.

Final Exam

(June 17, 2013)

1. (24 points) Determine whether each statement is true or not: True(T) or False(F). (If you answer correctly, you get 3 points; If nothing, 1 point; If wrong, 0 point.)
   1) The only ideals of a field $\mathbb{F}$ is $\{0\}$ and $\mathbb{F}$.
   2) The univariate polynomial ring $\mathbb{C}[x]$ is a principal ideal domain.
   3) Every UFD(unique factorization domain) is a PID.
   4) For a finite subset S of $\mathbb{C}[x,y]$, the zero set $V(S)$ is the same as $V(\langle S\rangle)$.
   5) There are many expressions for generating sets of an ideal of $\mathbb{C}[x,y]$.
   6) Every finite extension of fields is an algebraic extension.
   7) For a finite field $\mathbb{F}_p$ with p elements(p, a prime), there is an irreducible polynomial over $\mathbb{F}_p$ of any degree n.
   8) There exists a finite field with 10 elements.
2. (6pts) Among $\mathbb{Q}, \mathbb{Q}(\sqrt{2}), \mathbb{Q}(\sqrt{3}), \mathbb{Q}(\sqrt[3]{2}), \mathbb{R}, \mathbb{R}[x], \mathbb{C}, \mathbb{C}[x]$,
   1) $\mathbb{Q}[x]/\langle x^2-3\rangle$ is isomorphic to (   ).
   2) $\mathbb{R}[x]/\langle x^2+1\rangle$ is isomorphic to (   ).
3. (10pts) Let R be a commutative ring with 1, and I be an ideal of R. Then:
   1) R/I is an integral domain if and only if I is (   ) of R.
   2) R/I is a field if and only if I is (   ) of R.
4. (15pts) (1) Write the definition of a principal ideal domain.
   (2) State the Kronecker theorem on field extensions.
   (3) What is the definition of an algebraic extension of fields ?
5. (15pts) (1) Find $\text{irr}(\sqrt{2}+\sqrt{3}, \mathbb{Q})$.
   (2) Find $\text{irr}(1+\sqrt[3]{2}, \mathbb{Q})$.
   (3) Write all elements of $\mathbb{Q}(\sqrt[3]{2})$ using $\mathbb{Q}$.

• For each of the following problems, you should justify your answer.

6. (10pts) Show that every maximal ideal of a commutative ring with 1 is a prime ideal.
7. (20pts) For $p(x) \in \mathbb{F}[x]$, prove that $\langle p(x)\rangle$ is a maximal ideal of $\mathbb{F}[x]$ if and only if $p(x)$ is irreducible over $\mathbb{F}$.

Final Exam

(June 19, 2012)

1. (30 points) Determine whether each statement is true or not: True(T) or False(F). (If you answer correctly, you get 3 points; If nothing, 1 point; If wrong, 0 point.)
   1) The subring $\mathbb{Q}$ of $\mathbb{R}$ is an ideal of $\mathbb{R}$.
   2) For a ring homomorphism $\phi : R \to R'$ and an ideal I of R, $\phi(I)$ is an ideal of R'.
   3) A field has a nontrivial proper ideal.
   4) Every prime ideal of a commutative ring with 1 is a maximal ideal.
   5) Every ideal of $\mathbb{Z}[x]$ is principal.
   6) For an ideal I of a commutative ring R with 1, I is maximal in R if and only if R/I is a field.
   7) For each field F and a non-constant polynomial $f(x) \in F[x]$, there exists an extension field E of F such that $f(\alpha) = 0$ for some $\alpha \in E$.
   8) The element $\sqrt[3]{2} \in \mathbb{R}$ is algebraic over $\mathbb{Q}$.
   9) $\mathbb{Q}[\pi]$ is isomorphic to $\mathbb{Q}[x]$ as a ring.
   10) There exists a field with 10 elements.

2. (10pts) (1) Find an example of an integral domain which is not a field, except for $\mathbb{Z}$ and $\mathbb{F}[x]$ ($\mathbb{F}$, a field).
   (2) Find the multiplicative inverse of $x + \langle x^2 + 1 \rangle$ in the field $\mathbb{R}[x]/\langle x^2 + 1 \rangle$.
3. (10pts) (1) Write down the definition of a prime ideal P in a commutative ring R.
   (2) Find all prime ideals of $\mathbb{Z}$.
4. (15pts) Let $\mathbb{F}_3 = \{0, 1, 2\}$ be the field of 3 elements. Find an irreducible polynomial of degree 3 in $\mathbb{F}_3[x]$ and construct a field of 27 elements.

   • For each of the following problems, you should justify your answer.

5. (15pts) Show that $\mathbb{R}[x]/\langle x^2 + 1 \rangle$ is isomorphic to $\mathbb{C}$ as a ring.
6. (20pts) (1) Find $\operatorname{irr}(\sqrt{2} + \sqrt[4]{2}, \mathbb{Q})$. [Hint: $(x - \sqrt{2})^2 = \sqrt{2}$.]
   (2) Show that $\mathbb{Q}(\sqrt{2} + \sqrt[4]{2}) = \mathbb{Q}(\sqrt[4]{2})$.

4. Geometric constructions

DEFINITION 4.1. [6] A real number α is *constructible* if $|\alpha|$ can be constructed in a finite number of steps by a ruler with 1 and a compass. We can draw lines and circles in the plane $\mathbb{R}^2$.

EXERCISE 4.2. (1) Describe how to bisect the line segment.

(2) Describe how to erect the perpendicular line passing through a point on a line or outside a line.

(3) Describe how to find the parallel line passing through a point outside a line.

EXERCISE 4.3. (1) Construct a regular triangle, a square, and a regular hexagon.

(2) Show that bisecting the angle is possible, so the angle 30° is constructible.

(3) Show that the angle θ is constructible if and only if $\cos\theta$ is constructible.

Theorem 4.4. *The set F of all constructible real numbers forms a field.*

Proof. [7] For each nonzero $a, b \in F$, the real numbers $a \pm b, ab$ and b/a are also constructible. □

Corollary 4.5. *Note that $\mathbb{Q} \subseteq F$ since $1 \in F$. The points in $\mathbb{Q}^2$ are constructible.*

Our question is: Which real numbers are constructible?

[6]It is interesting to read articles `http://navercast.naver.com/science/math/935, -/1162, -/1299, -/1342, -/1501, -/7385.`

[7]René Descartes(1596–1650), a French philosopher.

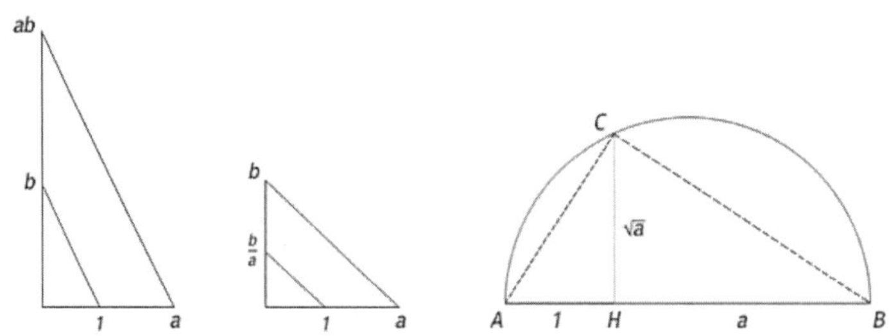

FIGURE 1. Construction of ab, b/a, and $\sqrt{a}$

Let L be any line between two constructible points, and C be a circle centered at a constructible point and passing through a constructible point. Then the intersection points of L and C are constructible. In other words, if H is a field of some constructible numbers, then a constructible number coming from the intersection points of L and C must lie in a field $H(\sqrt{\alpha})$, the smallest field containing H and $\sqrt{\alpha}$, for some $\alpha \in H$. If $\sqrt{\alpha} \notin H$, then $\sqrt{\alpha}$ is a zero of the irreducible polynomial $\mathrm{irr}(\sqrt{\alpha}, H) := x^2 - \alpha \in H[x]$. We denote the degree of $\mathrm{irr}(\sqrt{\alpha}, H)$ by $[H(\sqrt{\alpha}) : H] = 2$.

Moreover, if $\alpha \in F$ then $\sqrt{\alpha} \in F$ (See the figure below). For example, any rectangle whose sides are constructible is possible to be made into the square with the same area.

In summary, we conclude that:[8]

Theorem 4.6. $F = \{\gamma \in \mathbb{R} | \ \gamma \in \mathbb{Q}(\sqrt{\alpha_1}, \ldots, \sqrt{\alpha_n}),\ \alpha_1 \in \mathbb{Q}, \alpha_i \in \mathbb{Q}(\sqrt{\alpha_1}, \ldots, \sqrt{\alpha_{i-1}})\}$.

Corollary 4.7. *If $\gamma \in \mathbb{R}$ is a constructible number, then the degree of an irreducible polynomial of γ over $\mathbb{Q}$ must be a power of 2, that is,*

$$[\mathbb{Q}(\gamma) : \mathbb{Q}] = 2^n$$

for some integer $n \geq 0$. [9]

Now we answer the old three questions, presented by Greek, B.C. 4C.

Theorem 4.8 (Impossibility of certain constructions).

(a) *Doubling the cube is impossible.*

(b) *Squaring the circle is impossible.*

(c) *Trisecting the angle is impossible.*

Proof. (a) $\sqrt[3]{2} \notin F$.

(b) $\sqrt{\pi} \notin F$, since π is not an algebraic number.

(c) From $\cos 3\theta = 4\cos^3\theta - 3\cos\theta$, we obtain that $\cos 20°$ is a zero of an irreducible polynomial $8x^3 - 6x - 1 \in \mathbb{Q}[x]$. But $\deg(\cos 20°, \mathbb{Q}) = 3$ is not an even prime power, so $\cos 20°$ cannot be constructible. Therefore, the angle $60°$ cannot be trisected. □

[8]In 1837, by Pierre Wantzel(1814–1848), a French mathematician.

[9]The converse is not true. See [R, p.136].

Our question is:
Which regular n-gons are constructible? For example, could you construct a regular pentagon, regular heptagon, octagon, nonagon, decagon, undecagon, dodecagon?

Note that a regular n-gon can be drawn in the complex plane[10] by finding roots of the binomial equation

$$x^n = 1,$$

whose trivial root is $x = 1$.

Consider the binomial $x^p - 1$ (p, a prime). We know that

$$x^p - 1 = (x-1)(x^{p-1} + x^{p-2} + \cdots + x + 1),$$

and it is known that $x^{p-1} + x^{p-2} + \cdots + x + 1$ is irreducible over $\mathbb{Q}$.

DEFINITION 4.9. We define the p-th *cyclotomic polynomial* to be

$$\Phi_p(x) = x^{p-1} + x^{p-2} + \cdots + x + 1.$$

The degree $p-1$ of $\Phi_p(x)$ must be a power of 2 if a zero of $\Phi_p(x)$ is constructible in the complex plane($\cong \mathbb{R}^2$). Let $p - 1 = 2^k$. Suppose that $k = ab$ with $b > 1$, an odd integer. Then $2^k + 1 = (2^a + 1)(2^{(b-1)a} - 2^{(b-2)a} + \cdots - 2^a + 1)$ is not a prime. Thus, $k = 2^m$ for some nonnegative integer m.

DEFINITION 4.10. A prime of the form $2^{2^m} + 1$ is called a *Fermat prime.*

EXAMPLE 4.11. If $m = 0, 1, 2, 3, 4$, the Fermat primes $p = 2^{2^m} + 1$ are

$$3, \ 5, \ 17, \ 257, \ 65537.$$

Note that $2^{2^5} + 1$ is divisible by 641. In fact, it is not known what is the next Fermat prime.

Theorem 4.12. *Let p be a prime. Then a regular p-gon is constructible if and only if p is a Fermat prime.*

REMARK 4.13. * A regular 17-gon is constructible, since we have an explicit value[11]

$$\cos\frac{2\pi}{17} = \frac{1}{16}\left(-1+\sqrt{17}+\sqrt{34-2\sqrt{17}}+2\sqrt{17+3\sqrt{17}-\sqrt{34-2\sqrt{17}}-2\sqrt{34+2\sqrt{17}}}\right).$$

---

[10]It is called also as the Gauss plane. In 1796, he represents a complex number in the plane with the real and imaginary axes.

[11]Gauss calculated this when he was 18 years old.

In general, how about a regular n-gon, where n is not a prime?

EXERCISE 4.14. Show that a regular 9-gon is not constructible. [Hint: $60°$ cannot be trisected.]

Observe that among the nine roots of $x^9 - 1 = 0$, only 6 roots are essential for constructing a regular 9-gon. In fact, in order to construct a regular 9-gon, we have to find a zero of

$$\frac{x^9 - 1}{x^3 - 1} = x^6 + x^3 + 1.$$

Its degree is not a power of 2, so a regular 9-gon is not constructible. For a regular n-gon, what is the degree of the corresponding polynomial?

DEFINITION 4.15. The number of integers k such that $1 \le k < n$ and $\gcd(k, n) = 1$ is denoted by $\phi(n)$, which is called the *Euler phi function.*

EXAMPLE 4.16. (1) Let p be a prime. Then $\phi(p^e) = p^e - p^{e-1}$.

(2) For each m and n with $\gcd(m, n) = 1$, we get that $\phi(mn) = \phi(m)\phi(n)$.

Define the n-th cyclotomic polynomial to be

$$\Phi_n(x) := \prod_{i=1}^{\phi(n)} (x - \alpha_i),$$

where α_i's are primitive n-th roots of 1. Explicitly,

$$\{\alpha_i |\ 1 \le i \le \phi(n)\} = \{\zeta^k |\ 1 \le k < n,\ \gcd(k, n) = 1\},$$

with ζ, any primitive root of 1, say, $\zeta = \cos\frac{2\pi}{n} + i\sin\frac{2\pi}{n}$. It is known that $\Phi_n(x)$ is a polynomial with coefficients in $\mathbb{Q}$ and is an irreducible polynomial over $\mathbb{Q}$.

The degree $\phi(n)$ of $\Phi_n(x)$ must be a power of 2 if a zero of $\Phi_n(x)$ is constructible in the complex plane($\cong \mathbb{R}^2$).

Theorem 4.17 (Gauss). *A regular n-gon is constructible if and only if $n = 2^m$ or $n = 2^m p_1 \cdots p_r$, where p_i's are distinct Fermat primes.*

EXERCISE 4.18. Determine which of the following regular n-gons are constructible: $n = 7, 15, 20, 30, 36, 60$.

EXERCISE 4.19. (1) Prove that a regular 10-gon is constructible by showing $\cos\frac{2\pi}{10}$ is explicitly constructible. [Hint: Refer to Fig.2: the golden ratio φ is constructible.]

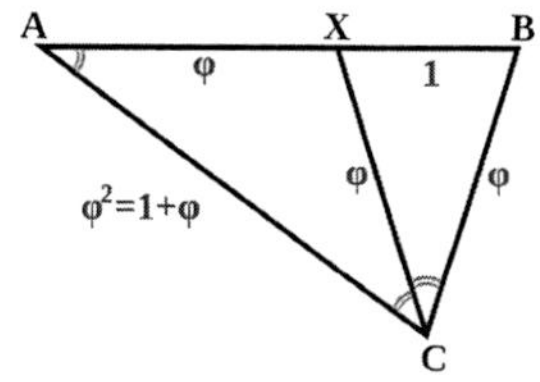

FIGURE 2. The angle $\frac{2\pi}{10}$

(2) Show that the angle 72° can be trisected. [Hint: $\cos\frac{\pi}{15}$ is constructible.] In general, for a positive integer n, the angle $n°$ is constructible if and only if n is a multiple of 3. Note that 1° is not constructible.

Final Exam

(June 17, 2010)

1. Determine whether each statement is true or not: True(T) or False(F).

   (If you answer correctly, you get 3 points; If nothing, 1 point; If wrong, 0 point.)

   1) Let D be a PID(principal ideal domain) which is not a field. Then an ideal $\langle p \rangle \leq D$ is maximal if and only if p is an irreducible.
   2) Every UFD(unique factorization domain) is a PID.
   3) If D is a UFD, then $D[x]$ is a UFD.
   4) The polynomial ring $\mathbb{F}[x, y]$ with two variables over a field $\mathbb{F}$ is a PID.
   5) $\mathbb{Z}[x]$ ($\mathbb{Z}$, the ring of integers) is a UFD.
   6) The field $\mathbb{R}[x]/\langle x^2 + 1 \rangle$ is isomorphic to the field $\mathbb{C}$ of complex numbers.
   7) Every nonconstant polynomial in $\mathbb{F}[x]$ has a zero in some extension field of $\mathbb{F}$.
   8) Every element of $\mathbb{C}$ is algebraic over $\mathbb{Q}$.
   9) Every constructible number is of degree 2^r over $\mathbb{Q}$ for some integer $r \geq 0$.
   10) It is possible to trisect the angle with a straightedge and a compass.

2. (20 pts) (1) Find $\operatorname{irr}(\sqrt{2+\sqrt{3}}, \mathbb{Q})$.

   (2) Find a basis and the dimension of $\mathbb{Q}(\sqrt{2+\sqrt{3}})$ as a vector space over $\mathbb{Q}$.

3. (20 pts) Prove that $\mathbb{Q}\left(\sqrt{3}+\sqrt{7}\right) = \mathbb{Q}\left(\sqrt{3}, \sqrt{7}\right)$.

4. (30 pts) Let $\mathbb{F}_2 = \{0, 1\}$ be the field of 2 elements, and $E = \mathbb{F}_2(\alpha)$ be the field of 2^3 elements, where α is a zero of $x^3 + x + 1$ in an extension field of $\mathbb{F}_2$. Complete the operation tables of E:

$+$	0	1	α	$1+\alpha$	α^2	$1+\alpha^2$	$\alpha+\alpha^2$	$1+\alpha+\alpha^2$
0	0	1	α	$1+\alpha$	α^2	$1+\alpha^2$	$\alpha+\alpha^2$	$1+\alpha+\alpha^2$
1								
α								
$1+\alpha$								
α^2								
$1+\alpha^2$								
$\alpha+\alpha^2$								
$1+\alpha+\alpha^2$								

,

$\times$	0	1	α	$1+\alpha$	α^2	$1+\alpha^2$	$\alpha+\alpha^2$	$1+\alpha+\alpha^2$
0	0	0	0	0	0	0	0	0
1	0	1	α	$1+\alpha$	α^2	$1+\alpha^2$	$\alpha+\alpha^2$	$1+\alpha+\alpha^2$
α								
$1+\alpha$								
α^2								
$1+\alpha^2$								
$\alpha+\alpha^2$								
$1+\alpha+\alpha^2$								

.

CHAPTER 7

Galois theory

Galois theory shows that there is a beautiful mapping,

$$\{ \text{ polynomials over a field } \} \longrightarrow \{ \text{ finite groups } \}.$$

More precisely, the extension field E containing the zeros of a polynomial $p(x) \in \mathbb{F}[x]$ is mapped into the group $\mathrm{Aut}_{\mathbb{F}}(E)$ of automorphisms of E permuting the zeros of $p(x)$ and fixing $\mathbb{F}$, and

$$\{ \text{ the subfields of } E \} \stackrel{1-1}{\longleftrightarrow} \{ \text{ the subgroups of } \mathrm{Aut}_{\mathbb{F}}(E) \}.$$

This chapter contains only brief materials mainly consisting of exercises and statements without proofs. For details, one may refer to the standard textbooks, e.g., the undergraduate textbook [F] or the graduate textbook [Hg].

1. Splitting fields

EXERCISE 1.1. Let F be an extension field of K and f a polynomial in $K[x]$. Show that if u is a root of f, then $\sigma(u)$ is also a root of f for all $\sigma \in \mathrm{Aut}_K F$.

DEFINITION 1.2. Let K e field and $f(x) \in K[x]$. Then the smallest extension field E of K containing all roots of $f(x)$ in an extension field is called the *splitting field* of $f(x)$. We say that E is a *normal* extension of K if E is a splitting field of some $f(x) \in K[x]$.

EXERCISE 1.3. Let f be a polynomial in $K[x]$ of degree n. Show that there exists a splitting field F of f over K with $[F : K] \leq n!$.

Proposition 1.4. *Let S be a set of polynomials in $K[x]$. If F is a splitting field of S over K and E is an intermediate field, then F is a splitting field of S over E.*

EXERCISE 1.5. Let α be a real number such that $\alpha^4 = 5$.

(1) Is $\mathbb{Q}(\alpha + i\alpha)$ normal over $\mathbb{Q}(i\alpha^2)$?

(2) Is $\mathbb{Q}(\alpha + i\alpha)$ normal over $\mathbb{Q}$? [Answer: $\alpha - i\alpha \notin \mathbb{Q}(\alpha + i\alpha)$.]

EXERCISE 1.6. (1) Prove or disprove: If F is normal over E and E is normal over K, then F is normal over K.

(2) Show that if F is an extension field of K such that $[F : K] = 2$, then F is a splitting field over K.

EXERCISE 1.7 (Isomorphism extension theorem). (1) Let $\sigma : K \to L$ be a field isomorphism, $S = \{f_i | \; i \in I\}$ a set of polynomials in $K[x]$, and $T = \{\sigma(f_i) | \; i \in I\} \subset L[x]$. Show that if F is a splitting field of S over K and M is a splitting field of T over L, then σ can be extended to an isomorphism $F \cong M$.

(2) Deduce that any two splitting fields of S over K are K-isomorphic. In particular, any two algebraic closures of K are K-isomorphic.

EXERCISE 1.8. (1) Show that F is an algebraic closure of K if and only if F is algebraic over K and F is algebraically closed.

(2) Let F be an algebraically closed field and E be the set of all elements in F that are algebraic over K. Show that E is an algebraic closure of K.

(3) Show that no finite field is algebraically closed.

2. Separable extensions

DEFINITION 2.1. (a) An irreducible polynomial $f \in K[x]$ is said to be *separable* if in some splitting field of f over K every root of f is a simple root.

(b) Let F be a extension field of K and $u \in F$ be algebraic over K. Then $u \in F$ is said to be *separable* over K if $\mathrm{irr}(u, K)$ is separable. If every element of F is separable over K, then F is said to be a *separable extension* of K.

EXERCISE 2.2. Let F be a field of characteristic 0. Show that every algebraic extension E of F is separable.

DEFINITION 2.3. A field F is called *perfect* if every algebraic extension of F is separable.

Theorem 2.4. *Let $K \leq E \leq F$ be extension fields. Then F is separable over K if and only if F is separable over E and E is separable over K.*

Theorem 2.5. *Let $K \leq E \leq F$ be extension fields and H be a subgroup of* $\mathrm{Aut}_K F$. *Then*

$$H' := \{v \in F | \; \sigma(v) = v \textit{ for all } \sigma \in H\}$$

is an intermediate field between K and F, which is called the fixed field of H. *Conversely.*

$$E' := \mathrm{Aut}_E F = \{\sigma \in \mathrm{Aut}_K F \mid \sigma(u) = u \textit{ for all } u \in E\}$$

is a subgroup of $\mathrm{Aut}_K F$, *called the* automorphism group of F fixing K.

REMARK 2.6. In the above theorem, clearly,

$$F' = \mathrm{Aut}_F F = \{\mathrm{id}\} =: 1, \ K' = \mathrm{Aut}_K F, \quad \text{and } 1' = F.$$

But it does not necessarily hold that $(\mathrm{Aut}_K F)' = K$.

DEFINITION 2.7. An extension field F of K is called a *Galois extension* if

$$(\mathrm{Aut}_K F)' = K.$$

Theorem 2.8. *Let F be an extension field of K. The following are equivalent*:

(i) *F is algebraic and Galois over K*;
(ii) *F is separable over K and F is a splitting field over K of a set of polynomials in $K[x]$*;
(iii) *F is a splitting field over K of a set of separable polynomials in $K[x]$.*

EXERCISE 2.9. Let F be an extension field of K. Show that if $u \in F$ is separable over K and $v \in F$ is purely inseparable over K then $K(u, v)$ is a simple extension of K. [Hint: Show that $K(u, v) = K(u + v)$.]

Theorem 2.10. *If F is a finite-dimensional separable extension of K, then F is a simple extension of K.*

Proof. The case of finite fields is OK. We assume that K is infinite. Since F is built up from a sequence of simple extensions, it suffices to consider the case $F = K(u, v)$.

Let $p(x), q(x) \in K[x]$ be the irreducible polynomials of u and v over K, and let the distinct roots of $p(x)$ and $q(x)$ in $\overline{K}$ be $u = u_1, \ldots, u_r$ and $v = v_1, \ldots, v_s$, respectively. By the separability assumption, $r = \deg p(x)$ and $s = \deg q(x)$. Since K is infinite, we can choose an element $t \in K$ for which $t \neq \frac{u - u_i}{v_j - v}$ whenever $j \neq 1$. Then taking $w = u + tv \in F$, we find that $w \neq u_i + tv_j$ whenever $j \neq 1$. Define the polynomial (of degree r)

$$h(x) = p(w - tx) \in K(w)[x] \subseteq F[x].$$

Then $h(v) = p(u) = 0$, but $h(v_j) \neq p(u_i) = 0$ for any $j \neq 1$ by construction of t, so none of the other v_j is a zero of $h(x)$. Now since the polynomials $h(x), q(x) \in K(w)[x]$ have exactly one common root in $\overline{K}$, namely v, by separability their greatest common divisor in $K(w)[x]$ is a linear polynomial which must be $x - v$, hence $v \in K(w)$ and so $u = w - tv \in K(w)$. This shows that $K(u, v) \leq K(w)$ and therefore $K(w) = K(u, v)$. □

EXERCISE 2.11. If $F = K(u, v)$ with u, v algebraic over K and v separable over K, then show that F is a simple extension of K.

EXERCISE 2.12. Let F be a finite extension field of K.

(1) Show that F is separable over K then there are only finitely many intermediate fields. [Hint: Galois theory]

(2) Show that if there are only finitely many intermediate fields then F is a simple extension of K. [Hint: Choose $u \in F$ such that $[K(u) : K]$ is maximal.]

Theorem 2.13 (Primitive element theorem; Artin[1]). *Let F be a finite extension field of K. Then F is a simple extension of K if and only if there are only finitely many intermediate fields.*

EXERCISE 2.14. Let F be a separable extension of a field K such that every element of F has degree less than a given positive integer n. Prove that $[F : K] < n$. [Hint: Primitive element theorem]

EXERCISE 2.15. Let $\operatorname{ch} K = p \neq 0$ and assume $F = K(u, v)$ where $u^p \in K, v^p \in K$ and $[F : K] = p^2$. Show that F is not an simple extension of K. Exhibit an infinite number of intermediate fields.

3. Fundamental theorem

Lemma 3.1. *Let F be an extension field of K with intermediate fields L and M. Let H and J be subgroups of $G = \mathrm{Aut}_K F$. Then,*

(a) *$F' = 1$ and $K' = G$, $1' = F$.*

(b) *$L < M$ implies that $M' < L'$. Conversely, $H < J$ implies that $J' < H'$.*

(c) *$L \leq L''$ and $H \leq H''$.*

(d) *$L' = L'''$ and $H' = H'''$.*

---

[1]Emil Artin(1898–1962), an Austrian-American mathematician.

Let F be an extension field of K. We denote $\mathrm{Aut}_K F$ by the *Galois group* of F over K.

REMARK 3.2. With the same notations as above,

(1) We know that F is a Galois extension of K if and only if $K = K''$.

(2) For X, an intermediate field or a subgroup of a Galois group $G = \mathrm{Aut}_K F$, we say that X is closed if $X = X''$.

Theorem 3.3. *Let F be an extension field of K. Then there is a 1-1 correspondence between the set of closed intermediate fields and the set of closed subgroups of* $\mathrm{Aut}_K F$.

EXERCISE 3.4. Let F be an extension field of K and E an intermediate field. Show that if E is an algebraic and Galois extension of K, then E is stable relative to F and K.

Theorem 3.5 (Fundamental theorem of Galois theory). *Let F be a finite-dimensional Galois extension of K. Then there is a 1-1 correspondence between the set of all intermediate fields and the set of all subgroups of* $\mathrm{Aut}_K F$ *satisfying*:

(i) $[\mathrm{Aut}_L F : \mathrm{Aut}_M F] = [M : L]$, *in particular*, $|\mathrm{Aut}_K F| = [F : K]$;

(ii) *For $K \leq E \leq F$, F is Galois over E. But E is Galois over K if and only if $E' = \mathrm{Aut}_E F$ is a normal subgroup of $\mathrm{Aut}_K F$. In this case,*

$$(\mathrm{Aut}_K F)/(\mathrm{Aut}_E F) \cong \mathrm{Aut}_K E.$$

Proof. For the full proof, consult [Hg, Ch.V §2]. □

Theorem 3.6 (Artin). *Let F be a field, G be a finite subgroup of* $\mathrm{Aut}(F)$, *and $K = G'$. Then F is a finite-dimensional Galois extension of K with $\mathrm{Aut}_K F = G$.*

Proof. Clearly, $G \leq \mathrm{Aut}_K F$. For $u \in F \setminus K$, there exists $\sigma \in G$ such that $\sigma(u) \neq u$ because $K = G'$, so $(\mathrm{Aut}_K F)' = K$, i.e., F/K is Galois. Since $|G| < \infty$, we get

$$[F : K] = [1' : G'] \leq [G : 1] = |G|,$$

i.e., F/K is finite-dimensional. Then K and G are closed, and we have $\mathrm{Aut}_K F = K' = G'' = G$. □

EXERCISE 3.7 (Elementary symmetric functions). (1) Let k be a field and $F = k(x_1, \ldots, x_n)$ be the field of rational functions in $x_1, \ldots, x_n$ over k. Construct a subfield K of F such that F is a Galois extension of K with Galois group S_n.

(2) Let G be a finite group. Find a Galois extension of fields whose Galois group is isomorphic to G. [Hint: Caley's theorem]

Solution. (1) The S_n-action on $k(x_1, \ldots, x_n)$ by $\sigma.x_i = x_{\sigma(i)}$ $(i = 1, \ldots, n)$ defines a k-automorphism of $k(x_1, \ldots, x_n)$. Then the map $S_n \to \mathrm{Aut}_k k(x_1, \ldots, x_n)$ is clearly a monomorphism of groups, whence S_n may be considered to be a subgroup of $\mathrm{Aut}_k k(x_1, \ldots, x_n)$. Let $K = S_n'$ be the field of all symmetric functions in $x_1, \ldots, x_n$ over k. By Artin's theorem, $k(x_1, \ldots, x_n)/K$ is a Galois extension with Galois group S_n.

For the construction of K, let $f_1, f_2, \ldots, f_n$ be the elementary symmetric functions in $x_1, \ldots, x_n$ over k. Clearly, $k(f_1, \ldots, f_n) \le K$. We show $K = k(f_1, \ldots, f_n)$. Since

$$[k(x_1, \ldots, x_n) : K] = |S_n| = n!,$$

it suffices to show $[k(x_1, \ldots, x_n) : k(f_1, \ldots, f_n)] \le n!$. Let $K_0 = k(f_1, \ldots, f_n)$ and consider

$$K_0 \le K_0(x_n) \le K_0(x_{n-1}, x_n) \le \cdots \le K_0(x_1, \ldots, x_{n-1}, x_n) = k(x_1, \ldots, x_n).$$

It suffices to show that $[K_0(x_n) : K_0] \le n$ and

$$[(K_0(x_{k+1}, \ldots, x_n))(x_k) : K_0(x_{k+1}, \ldots, x_n)] \le k$$

for each $k < n$. Let

$$g_n(y) = (y - x_1)(y - x_2) \cdots (y - x_n) = y^n - f_1 y^{n-1} + \cdots + (-1)^n f_n \in K_0[y].$$

Then $[K_0(x_n) : K_0] = \deg(x_n, K_0) \le \deg g_n = n$. For each $k < n$, let

$$g_k(y) = (y - x_1)(y - x_2) \cdots (y - x_k).$$

Then $g_k(y) \in K_0(x_{k+1}, \ldots, x_n)[y]$ by induction on k, whence x_k is algebraic of degree at most k over $K_0(x_{k+1}, \ldots, x_n)$.

We've proved that $k(x_1, \ldots, x_n)/k(f_1, \ldots, f_n)$ is a Galois extension with Galois group S_n.

(2) By Cayley's theorem, a finite group G is isomorphic to a subgroup of S_n (also denoted by G). Then, for the fixed field E of G, $k(x_1, \ldots, x_n)/E$ is a Galois extension with $\mathrm{Aut}_E k(x_1, \ldots, x_n) = G$ by Galois theory. □

Example 3.8. Let K be the splitting field of $f(x) = x^3 - 2$ over $\mathbb{Q}$. Then $\sqrt[3]{2}$, $\sqrt[3]{2}\omega$, $\sqrt[3]{2}\omega^2$ ($\omega = \frac{-1+\sqrt{3}i}{2}$, a primitive 3rd root of 1) are all zeros of $f(x)$. Then

$[\mathbb{Q}(\sqrt[3]{2}) : \mathbb{Q}] = 3$. Since $[K : \mathbb{Q}] > 3$, we obtain that $[K : \mathbb{Q}] = 3! = 6$ and $\mathrm{Aut}_{\mathbb{Q}}(K) \cong S_3$. In fact, $K = \mathbb{Q}(\sqrt[3]{2}, \sqrt{3}i)$.

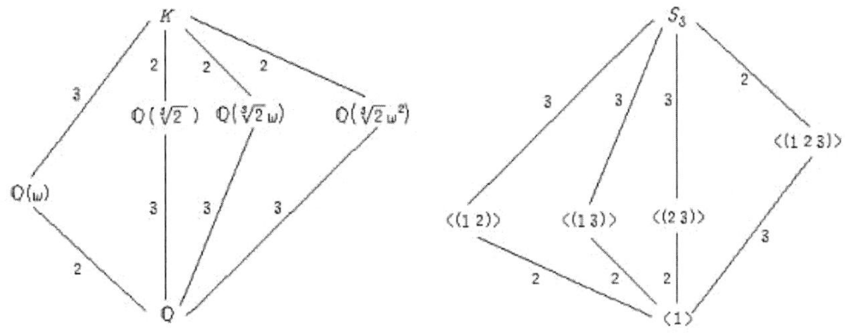

FIGURE 1. Correspondence of $x^3 - 2 \in \mathbb{Q}[x]$ between fixed fields and automorphism groups

The following exercises deal with the transcendental extension field $K(x)$ over K.

EXERCISE 3.9. Show that $\mathrm{Aut}_{\mathbb{Q}}\mathbb{R}$ is trivial. [Hint: Since each $r \in \mathbb{R}_+$ is a square, any $\sigma \in \mathrm{Aut}\mathbb{R}$ sends $\mathbb{R}_+$ to $\mathbb{R}_+$, i.e., σ preserves the order in $\mathbb{R}$.]

EXERCISE 3.10. Let $K(x)/K$ be a field extension and $f/g \in K(x) \setminus K$ with $f, g \in K[x]$ relatively prime. Show the following:

(1) x is algebraic over $K(f/g)$ and $[K(x) : K(f/g)] = \max(\deg f, \deg g)$. So if $E \neq K$ is an intermediate field, then $[K(x) : E] < \infty$.

(2) The assignment $x \mapsto f/g$ induces a homomorphism $\sigma : K(x) \to K(x)$ such that $\varphi(x)/\psi(x) \to \varphi(f/g)/\psi(f/g)$. σ is a K-automorphism of $K(x)$ if and only if $\max(\deg f, \deg g) = 1$. So $\mathrm{Aut}_K K(x)$ consists of all those automorphisms induced by the assignment

$$x \mapsto (ax + b)/(cx + d)$$

where $a, b, c, d \in K$ and $ad - bc \neq 0$.

EXERCISE 3.11. Let $G_1 = \{\sigma_i \in \mathrm{Aut}_K K(x) |\ \sigma_i(x) = x^i,\ i = \pm 1\}$ and

$$G_2 = \{\sigma_1, \sigma_2, \sigma_3 \in \mathrm{Aut}_K K(x) |\ \sigma_1(x) = x, \sigma_2(x) = 1/(1 - x), \sigma_3(x) = (x - 1)/x\}.$$

Then G_1, G_2 are subgroups of $\mathrm{Aut}_K K(x)$. Determine the fixed fields of G_1, G_2. [Answer: $G_1' = K(x + x^{-1})$.]

EXERCISE 3.12. (1) Show that if K is an infinite field, then $K(x)$ is Galois over K.

(2) Show that if K is finite, then $K(x)$ is not Galois over K.

Solution. (1) Suppose that $K(x)/K$ is not Galois, i.e., $E := (\mathrm{Aut}_K K(x))' \gneq K$. Then $[K(x) : E] < \infty$. But $\mathrm{Aut}_E K(x) = \mathrm{Aut}_K K(x)$ is infinite since there are infinitely many assignments of the form $x \mapsto (ax+b)/(cx+d)$, which contradicts

$$[\mathrm{Aut}_K K(x) : 1] \leq [K(x) : E].$$

(2) $\mathrm{Aut}_K K(x)$ is finite. Suppose that $K(x)/K$ is Galois. Then

$$[K(x) : K] \leq [\mathrm{Aut}_K K(x) : 1],$$

which is a contradiction. □

EXERCISE 3.13. Show the following:

(1) Assume $\operatorname{ch} K = 0$ and $G \leq \mathrm{Aut}_K K(x)$ be a subgroup generated by the automorphism induced by $x \mapsto x+1$. Then G is an infinite cyclic group, and $G' = K$.

(2) If K is an infinite field, then the only closed subgroup of $\mathrm{Aut}_K K(x)$ are itself and its finite subgroups.

(3) In the extension of $\mathbb{Q}$ by $\mathbb{Q}(x)$, the intermediate field $\mathbb{Q}(x^2)$ is closed, but $\mathbb{Q}(x^3)$ is not.

EXERCISE 3.14. In the extension of an infinite field K over $K(x, y)$, show that the intermediate field $K(x)$ is Galois over K, but not stable relative to $K(x, y)$ and K.

EXERCISE 3.15. Let E, F be finite Galois extension of a field K such that the K-algebra $E \otimes_K F$ is an integral domain.

(1) Prove that $E \otimes_K F$ is a field.

[Hint: Let $n = \dim_K(E \otimes_K F) \leq (\dim_K E)(\dim_K F)$.]

(2) Is $E \otimes_K F$ a Galois extension of K?

EXERCISE 3.16. Let F be a finite extension field of a finite field K. Show that F/K is Galois and the Galois group of F/K is cyclic.

4. The Galois group of a polynomial

Let E be an extension field of F. We denote $\mathrm{Aut}_F E$ by the Galois group $Gal(E/F)$. The Galois group of $f(x) \in F[x]$ is $Gal(E/F)$ where E is the splitting field of $f(x)$.

EXERCISE 4.1 (Discriminant). Let k be a perfect field and let $f(x)$ be a polynomial over k.

(1) Give a definition of the discriminant D of $f(x)$.

(2) Show that $D = \Delta^2$ is contained in k.

EXERCISE 4.2. Let K be the splitting field of $x^3 - 5$ over $\mathbb{Q}$

(1) Find $Gal(K/\mathbb{Q})$ up to isomorphism. [Hint: discriminant or cyclotomic extension]

(2) Find the number of all distinct subfields of K.

EXERCISE 4.3. Let $f(x) = x^3 - x - 2 \in \mathbb{Q}[x]$

(1) Find the Galois group of $f(x)$ over $\mathbb{Q}$, up to isomorphism.

(2) Find the Galois group of $f(x)$ over $\mathbb{Q}(\sqrt{-11})$, up to isomorphism.

EXERCISE 4.4. Find the Galois group of $x^3 + x + 1$ over F for each of the following cases:

(1) $F = \mathbb{C}$, (2) $F = \mathbb{R}$, (3) $F = \mathbb{Q}$, (4) $F = \mathbb{F}_3$, (5) $F = \mathbb{F}_5$.

[Hint: Check irreducibility and discriminant.]

DEFINITION 4.5. An extension field $E \geq F$ is an *abelian extension* if $E \geq F$ is a Galois extension whose Galois group is abelian.

EXERCISE 4.6. Let $f_p(x) = x^3 - px + 2p \in \mathbb{Z}[x]$. Find all primes p less than 100 for which the splitting field of $f_p(x)$ is an abelian extension over $\mathbb{Q}$.

EXERCISE 4.7. (1) Let $f(x) = x^4 - 5$ and F be a splitting field of $f(x)$ over $\mathbb{Q}$. Describe the Galois correspondence between the intermediate subfields of F and the subgroups of the Galois group of $f(x)$ over $\mathbb{Q}$.

(2) Show that the Galois group of the polynomial of $x^4 - 5$ over $\mathbb{Q}(\zeta_5)$ is isomorphic to a subgroup of the Klein four group, where ζ_5 is a primitive 5-th root of 1 in $\overline{\mathbb{Q}}$.

EXERCISE 4.8. Let K/F be a Galois extension with the Galois group isomorphic to the alternating group A_4. Show that there exists a unique Galois extension E of F contained in K such that $[E : F] = 3$.

5. Cyclotomic extensions

DEFINITION 5.1. Let n be a positive integer, and α_i's be primitive n-th roots of unity. Then we define the n-th *cyclotomic polynomial* to be

$$\Phi_n(x) := \prod_{i=1}^{\phi(n)} (x - \alpha_i).$$

EXERCISE 5.2. Let $F \geq \mathbb{Q}$ be the splitting field of $\Phi_n(x)$, i.e., the cyclotomic extension over $\mathbb{Q}$ of order n. Show that

(1) $\Phi_n(x)$ is irreducible in $\mathbb{Q}[x]$, (2) $[F : \mathbb{Q}] = \phi(n)$, (3) $\mathrm{Aut}_{\mathbb{Q}}F \cong \mathbb{Z}_n^{\times}$.

EXERCISE 5.3. Let p be an odd prime and ζ_p a primitive p-th root of 1.

(1) Find $[\mathbb{Q}(\zeta_p) : \mathbb{Q}(\zeta_p + \zeta_p^{-1})]$.

(2) Show that $\mathbb{Q}(\zeta_p + \zeta_p^{-1}) = \mathbb{R} \cap \mathbb{Q}(\zeta_p)$.

EXERCISE 5.4. Let $\zeta = e^{2\pi i/11}$.

(1) Describe the following Galois groups up to isomorphism:

1) $Gal(\mathbb{Q}(\zeta)/\mathbb{Q})$, 2) $Gal(\mathbb{Q}(\zeta)/\mathbb{Q}(\zeta + \zeta^{-1}))$, 3) $Gal(\mathbb{Q}(\zeta + \zeta^{-1})/\mathbb{Q})$.

(2) Find the unique quadratic field L over $\mathbb{Q}$ contained in $\mathbb{Q}(\zeta)$.

EXERCISE 5.5. Let $f(x) = x^6 + x^3 + 1$.

(1) Show that $f(x)$ is irreducible over $\mathbb{Q}$.

(2) Find the Galois group of the splitting field F of $f(x)$ over $\mathbb{Q}$ up to isomorphism. [Answer: $\mathrm{Aut}_{\mathbb{Q}}\mathbb{Q}(e^{2\pi i/9}) \cong (\mathbb{Z}/9\mathbb{Z})^{\times} \cong \mathbb{Z}/6\mathbb{Z}$.]

EXERCISE 5.6. Let K be a splitting field of $f(x) = x^5 - 2$ over $\mathbb{Q}$.

(1) Find $H = Gal(K/\mathbb{Q}(\zeta))$, where ζ is a primitive 5-th root of 1. [Answer: $H = \{\sigma^i |\ i = 0, 1, 2, 3, 4\}$ where $\sigma(\sqrt[5]{2}) = \sqrt[5]{2}\,\zeta$.]

(2) Find a solvable series of $G = Gal(K/\mathbb{Q})$. [Answer: $G \rhd D_{2\cdot 5} \rhd \mathbb{Z}/5\mathbb{Z} \rhd 0$.]

EXERCISE 5.7. Let k be a field and ζ_n a primitive n-th root of 1 in $\overline{k}$.

(1) Show that $Gal(k(\zeta_n)/k)$ is isomorphic to a subgroup of $(\mathbb{Z}/n\mathbb{Z})^{\times}$.

(2) Show that every cyclotomic extension is an abelian extension.

Note that $\mathbb{Q} < \mathbb{Q}\left(\zeta + \frac{1}{\zeta}\right) < \mathbb{Q}(\zeta)$ and $\left[\mathbb{Q}\left(\zeta + \frac{1}{\zeta}\right) : \mathbb{Q}\right] = \frac{\phi(n)}{2}$. So, a regular n-gon is constructible only if $\phi(n) = 2^k$.

DEFINITION 5.8. A prime of the form $p = 2^{2^m} + 1$ is called a *Fermat prime*.

EXAMPLE 5.9. If $m = 0, 1, 2, 3, 4$, the Fermat primes $p = 2^{2^m} + 1$ are

$$3,\ 5,\ 17,\ 257,\ 65537.$$

Note that $2^{2^5}+1$ is divisible by 641. In fact, it is not known what is the next Fermat prime.

Theorem 5.10 (Gauss). *A regular n-gon is constructible if and only if an odd prime factor p of n is a Fermat prime such that $p^2 \nmid n$.*

6. Cyclic extensions *

EXERCISE 6.1 (Trace & Norm). Let E/F be a finite Galois extension.

(1) Show that $\mathrm{tr}_{E/F}$ and $\mathrm{N}_{E/F}$ maps E into F.

(2) Assume that F is the quotient field of an integral domain D. Show that if $\alpha \in E$ is integral over D then $\mathrm{tr}_{E/F}(\alpha)$ and $\mathrm{N}_{E/F}(\alpha)$ are integral over D.

(3) Assume that F is a field of characteristic 0. Show that $\mathrm{tr}_{E/F} : E \to F$ is surjective.

(4) Assume that F is a finite field. Show that $\mathrm{N}_{E/F} : E \to F$ is surjective.

DEFINITION 6.2. An extension field $E \geq F$ is an *cyclic extension* if $E \geq F$ is a Galois extension whose Galois group is cyclic.

EXERCISE 6.3. Let E be a cyclic extension of F of degree n and $G = \mathrm{Aut}_F E = \langle \sigma \rangle$. Show that for $u \in E$, $N_{E/F}(u) = 1$ if and only if $u = v\sigma(v)^{-1}$ for some $v \in E$.

EXERCISE 6.4. Let F be a field which contains a primitive n-th root of unity ζ_n. Show that if E is a cyclic extension field over F of degree n then E is a splitting field over F of an irreducible polynomial of the form $x^n - b \in F[x]$.

EXERCISE 6.5. Let $\alpha = \overline{\mathbb{Q}} \setminus \mathbb{Q}$, where $\overline{\mathbb{Q}}$ is an algebraic closure of $\mathbb{Q}$.

(1) Show that there exists a subfield F of $\overline{\mathbb{Q}}$ containing $\mathbb{Q}$ such that F is maximal with respect to the property $\alpha \notin F$.

(2) Prove that every Galois extension E/F is cyclic, where $E \leq \overline{\mathbb{Q}}$.

7. Radical extensions

It has been known how to explicitly solve polynomial equations

- of degree 2 (Babylonians, many centuries ago),
- of degree 3 (Scipione del Ferro, Tartaglia, Cardano, around 1500), and

- even of degree 4 (Cardano, Ferrari, 16th century), using only algebraic operations $(+, -, \times, \div)$ and radicals (nth roots).

- However, the case of degree 5 remained unsolved until Abel showed in 1824 that a general equation of degree 5 cannot be solved that way.

In 1770, Lagrange doubts that there is a root formula for a quintic equation and he thinks it's crucial to understand the symmetric group S_n. At that time, mathematics is not mature yet. For instance, even Euler was wrong in that $\sqrt{-2}\cdot\sqrt{-3} = \sqrt{6}$, and Gauss don't like using imaginary numbers just after the notion of complex numbers, even though he himself proves around 1800 that any polynomial in $\mathbb{C}$ has a zero in $\mathbb{C}$.

Definition 7.1. (a) An extension field $K \geq F$ is an *extension by radicals* if there exist $\alpha_1, \alpha_2, \ldots, \alpha_r \in \overline{F}$ and positive integers $n_1, n_2, \ldots, n_r$ such that

$$K = F(\alpha_1, \alpha_2, \ldots, \alpha_r),\ \alpha_1^{n_1} \in F,\ \alpha_i^{n_i} \in F(\alpha_1, \ldots, \alpha_{i-1})$$

for $1 < i \leq r$.

(b) We say that $f(x) \in F[x]$ is *solvable by radicals* over F if the splitting field E of $f(x)$ is contained in an extension field by radicals.

Example 7.2. (1) $x^5 - 1$ is solvable by radicals over $\mathbb{Q}$, since we have the splitting field $E = \mathbb{Q}(\zeta)$ where ζ is a primitive root of 1, satisfying $\zeta^5 = 1 \in \mathbb{Q}$.

(2) $x^5 - 2$ is also solvable by radicals over $\mathbb{Q}$, since we have the splitting field

$$E = \mathbb{Q}(\sqrt[5]{2}, \zeta).$$

Assume that, for all fields F, $\operatorname{ch} F = 0$.

Lemma 7.3. *Let $a \in F$ and E be the splitting field of $x^n - a \in F[x]$. Then* $\mathrm{Aut}_F E$ *is a solvable group.*

Exercise 7.4. Let K be a Galois and radical extension of F. Thus $K = F(u_1, \ldots, u_n)$, where $u_1^{m_1} \in F$, $u_i^{m_i} \in F(u_1, \ldots, u_{i-1})$ for $i \geq 2$. Assume that $\operatorname{ch} F$ does not divide m_i. Set $m = m_1 \cdots m_n$ and let ζ be a primitive m-th root of unity.

(1) Show that $\mathrm{Aut}_{F(\zeta)} K(\zeta)$ is solvable.

(2) Deduce that $\mathrm{Aut}_F K(\zeta)$ is solvable.

(3) Conclude that $\mathrm{Aut}_F K$ is solvable.

(4) Let $f \in F[x]$. Show that if $f(x) = 0$ is solvable by radicals then the Galois group of f is solvable.

Corollary 7.5 (Abel 1824). *A general equation of degree $n \geq 5$ over a field F is not solvable by radicals.*

EXAMPLE 7.6. Consider $f(x) = x^5 - 10x + 5 \in \mathbb{Q}[x]$ and let E be the splitting field of $f(x)$. Note that $f(x)$ is irreducible over $\mathbb{Q}$ and it has three real zeros and two nonreal complex zeros. If α is a zero of $f(x)$ then $[\mathbb{Q}(\alpha) : \mathbb{Q}] = 5$. Thus the order of $\mathrm{Aut}_{\mathbb{Q}}(E)$ is divisible by 5, and the group contains a 5-cycle. Since the complex conjugation is an automorphism of E, $\mathrm{Aut}_{\mathbb{Q}}(E)$ contains a transposition. Therefore, the Galois group of $f(x)$ is S_5, which is not solvable, and so $f(x)$ is not solvable by radicals.

EXERCISE 7.7. Let $f(x) = x^5 - 16x + 2 \in \mathbb{Q}[x]$.

(1) Show that $f(x)$ is irreducible over $\mathbb{Q}$ having three real roots.

(2) Is $f(x)$ solvable by radicals?

EXERCISE 7.8. (a) Show that $f(x) = x^5 - 4x + 2 \in \mathbb{Q}[x]$ is not solvable by radicals.

(2) What is the Galois group of $x^5 - 2 \in \mathbb{Q}[x]$?

DEFINITION 7.9 (Distinguished class of extension fields). * A class $\mathcal{C}$ of extension fields $F \leq E$ is called *distinguished* if it satisfies the following conditions:

(1) Let $k \leq F \leq E$ be a tower of fields. Then $(k \leq E) \in \mathcal{C} \Leftrightarrow (k \leq F) \in \mathcal{C}$ and $(F \leq E) \in \mathcal{C}$.

(2) If $(k \leq E) \in \mathcal{C}$, if $k \leq F$ is any extension, and E, F are both contained in some field, then $(F \leq EF) \in \mathcal{C}$.

(3) If $(k \leq F), (k \leq E) \in \mathcal{C}$, and F, E are subfields of a common fields, then $(k \leq FE) \in \mathcal{C}$.

REMARK 7.10. Condition (3) is implied by (1) and (2).

EXAMPLE 7.11. * (1) The class of algebraic extensions is distinguished, and so is the class of finite extensions.

(2) The class of normal extensions is *not* distinguished.

(3) Separable extensions form a distinguished class of extensions.

(4) Purely inseparable extensions form a distinguished class of extensions.

(5) Solvable extensions form a distinguished class of extensions.

(6) Finitely generated extensions form a distinguished class of extensions.

8. Algebraic equations

For a quadratic equation

$$x^2 + ax + b = 0 \quad (a, b \in \mathbb{Q}),$$

we know the root formula:

$$x = \frac{-a \pm \sqrt{a^2 - 4b}}{2}.$$

Let F be the splitting field of $x^2 + ax + b \in \mathbb{Q}[x]$, that is, $F = \mathbb{Q}(\sqrt{a^2 - 4b})$. We note that, if the determinant $D = a^2 - 4b$ is not a square of a rational number, then

$$\mathrm{Aut}_{\mathbb{Q}}(F) = \{\mathrm{id.}, \sigma\} \cong \mathbb{Z}_2,$$

where $\sigma(\sqrt{a^2 - 4b}) = -\sqrt{a^2 - 4b}$.

8.1. Cubic equations.

How to solve a cubic equation $x^3 + ax^2 + bx + c = 0$ $(a, b, c \in \mathbb{Q})$? By substituting $y = x + \frac{a}{3}$, the equation is transformed into a simpler form $y^3 + py + q = 0$ $(p, q \in \mathbb{Q})$. Now we solve

$$x^3 + px + q = 0 \quad (p, q \in \mathbb{Q}).$$

Let D be the discriminant of $x^3 + px + q$, that is, $D = -(4p^3 + 27q^2)$, and let $\omega = \frac{-1+\sqrt{-3}}{2}$ which satisfies $\omega^3 = 1, \omega \neq 1$, i.e., $\omega^2 + \omega + 1 = 0$.

If we let α, β, γ be the three solutions of $x^3 + px + q = 0$, then, from $x^3 + px + q = (x - \alpha)(x - \beta)(x - \gamma)$, we have

$$\begin{cases} \alpha + \beta + \gamma = 0, \\ \alpha\beta + \beta\gamma + \gamma\alpha = p, \\ \alpha\beta\gamma = -q. \end{cases}$$

We set θ_1, θ_2 satisfying

$$\begin{cases} \alpha = \theta_1 + \theta_2, \\ \beta = \omega\theta_1 + \omega^2\theta_2, \\ \gamma = \omega^2\theta_1 + \omega\theta_2, \end{cases}$$

so that

$$3\theta_1 = \alpha + \omega^2\beta + \omega\gamma, \quad 3\theta_2 = \alpha + \omega\beta + \omega^2\gamma.$$

Then,

$$(3\theta_1)(3\theta_2) = (\alpha + \beta + \gamma)^2 - 3(\alpha\beta + \beta\gamma + \gamma\alpha) = -3p,$$

thus,

$$\theta_1\theta_2 = -\frac{p}{3}.$$

On the other hand,

$$(\theta_1+\theta_2)^3 = \begin{cases} \alpha^3 - p\alpha - q \\ \theta_1^3+\theta_2^3+3\theta_1\theta_2(\theta_1+\theta_2) = \theta_1^3+\theta_2^3 - p\alpha \end{cases},$$

so

$$\theta_1^3+\theta_2^3 = -q.$$

Hence θ_1^3 and θ_2^3 are the solutions of the quadratic equation

$$x^2+qx-\frac{p^3}{27}=0.$$

We calculate

$$\theta_1^3 = -\frac{q}{2}+\frac{\sqrt{-3D}}{18}, \quad \theta_2^3 = -\frac{q}{2}-\frac{\sqrt{-3D}}{18}.$$

Once we fix

$$\theta_1 = \sqrt[3]{-\frac{q}{2}+\frac{\sqrt{-3D}}{18}}, \quad \theta_2 = \sqrt[3]{-\frac{q}{2}-\frac{\sqrt{-3D}}{18}}$$

such that $\theta_1\theta_2 = -\frac{p}{3}$, then our three solutions α, β, γ follow.

Remark 8.1. Let F be the splitting field of $x^3+px+q \in \mathbb{Q}[x]$. Then, for a general cubic equation, we know that

$$\mathrm{Aut}_{\mathbb{Q}}(F) \cong \langle (1,2,3),(2,3)\rangle = S_3.$$

8.2. Quartic equations.

How to solve a quartic equation

$$x^4+px^2+qx+r=0 \quad (p,q,r\in\mathbb{Q})?$$

Let $\alpha_1,\alpha_2,\alpha_3,\alpha_4$ be 4 roots of $x^4+px^2+qx+r=0$, and consider

$$\begin{cases} \theta_1 = (\alpha_1+\alpha_2)(\alpha_3+\alpha_4), \\ \theta_2 = (\alpha_1+\alpha_3)(\alpha_2+\alpha_4), \\ \theta_3 = (\alpha_1+\alpha_4)(\alpha_2+\alpha_3). \end{cases}$$

Then $\theta_1 = (\alpha_1+\alpha_2)(\alpha_3+\alpha_4)$ together with $\alpha_1+\alpha_2+\alpha_3+\alpha_4=0$ implies that

$$\alpha_1+\alpha_2 = \sqrt{-\theta_1}, \ \alpha_3+\alpha_4 = -\sqrt{-\theta_1}.$$

In the same way,

$$\alpha_1+\alpha_3=\sqrt{-\theta_2},\ \alpha_2+\alpha_4=-\sqrt{-\theta_2},$$

$$\alpha_1+\alpha_4=\sqrt{-\theta_3},\ \alpha_2+\alpha_3=-\sqrt{-\theta_3},$$

where it has to be satisfied that

$$\sqrt{-\theta_1}\sqrt{-\theta_2}\sqrt{-\theta_3}=(\alpha_1+\alpha_2)(\alpha_1+\alpha_3)(\alpha_1+\alpha_4)$$

$$=\alpha_1^3+\alpha_1^2(\alpha_2+\alpha_3+\alpha_4)+\alpha_1(\alpha_2\alpha_3+\alpha_2\alpha_4+\alpha_3\alpha_4)+\alpha_2\alpha_3\alpha_4=-q.$$

From $\sqrt{-\theta_1}+\sqrt{-\theta_2}+\sqrt{-\theta_3}=3\alpha_1+\alpha_2+\alpha_3+\alpha_4=2\alpha_1$ and its analogues,

$$\begin{cases}\alpha_1=\frac{1}{2}\left(\sqrt{-\theta_1}+\sqrt{-\theta_2}+\sqrt{-\theta_3}\right),\\ \alpha_2=\frac{1}{2}\left(\sqrt{-\theta_1}-\sqrt{-\theta_2}-\sqrt{-\theta_3}\right),\\ \alpha_3=\frac{1}{2}\left(-\sqrt{-\theta_1}+\sqrt{-\theta_2}-\sqrt{-\theta_3}\right),\\ \alpha_4=\frac{1}{2}\left(-\sqrt{-\theta_1}-\sqrt{-\theta_2}+\sqrt{-\theta_3}\right).\end{cases}$$

Hence solving a quartic equation reduces to finding the solutions $\theta_1,\theta_2,\theta_3$:

$$\begin{cases}\theta_1+\theta_2+\theta_3=2\sum_{1\le i<j\le 4}\alpha_i\alpha_j=2p,\\ \theta_1\theta_2+\theta_2\theta_3+\theta_1\theta_3=p^2-4r,\\ \theta_1\theta_2\theta_3=-q^2.\end{cases}$$

Thus, $\theta_1,\theta_2,\theta_3$ are the solutions of the cubic equation

$$x^3-2px+(p^2-4r)x+q^2=0.$$

Obtaining these three solutions, we get $\alpha_1,\alpha_2,\alpha_3,\alpha_4$.

Exercise 8.2. In the calculation above, show that

$$\theta_1\theta_2=\alpha_1^2(\alpha_2\alpha_3+\alpha_3\alpha_4+\alpha_2\alpha_4)+\alpha_2^2(\alpha_1\alpha_3+\alpha_1\alpha_4+\alpha_3\alpha_4)$$

$$+\alpha_3^2(\alpha_1\alpha_2+\alpha_2\alpha_4+\alpha_1\alpha_4)+\alpha_4^2(\alpha_1\alpha_2+\alpha_2\alpha_3+\alpha_1\alpha_3)+\alpha_1^2\alpha_4^2+\alpha_2^2\alpha_3^2+2\alpha_1\alpha_2\alpha_3\alpha_4,$$

and

$$\theta_1\theta_2+\theta_2\theta_3+\theta_1\theta_3=(\sum_{1\le i<j\le 4}\alpha_i\alpha_j)^2+(\sum_{1\le i\le 4}\alpha_i)(\sum_{1\le i<j<k\le 4}\alpha_i\alpha_j\alpha_k)-4\alpha_1\alpha_2\alpha_3\alpha_4.$$

Remark 8.3. Let F be the splitting field of $x^4+px^2+qx+r\in\mathbb{Q}[x]$. Then, for a general quartic equation, we know that

$$\mathrm{Aut}_{\mathbb{Q}}(F)\cong\langle(1,2,3,4),(1,2)\rangle=S_4.$$

Remark 8.4 (Abel 1824). For a general quintic equation, there is no algebraic way to solve a quintic equation, since its Galois group is S_5 which is not a solvable group.

9. Transcendental extensions *

EXERCISE 9.1 (Transcendence base). Let E/K be a field extension.

(1) Give a definition of an algebraically independent subset in E over K.

(2) Show that there exists a transcendence base of E over K.

REMARK 9.2. The cardinality of a transcendence base of E over K is invariant, so the *transcendence degree* of E over K is defined to be the cardinality of any transcendence base of E over K, denoted by tr.d.E/K.

Theorem 9.3. *Let $K \leq E \leq F$ be extension fields. Then*

$$\text{tr.d.}F/K = \text{tr.d.}F/E + \text{tr.d.}E/K.$$

EXERCISE 9.4. Let F be a finitely generated extension of K and E be an intermediate field. Then show that E is finitely generated over K.

DEFINITION 9.5. Let C be an algebraically closed field with subfields K, E, F such that $K \leq E \cap F$. E and F are *linearly disjoint* over K if every subset of E which is linearly independent over K is also linearly independent over F.

REMARK 9.6. The definition of linear disjointness is symmetric in E and F.

EXERCISE 9.7. If E is a purely transcendental extension of K and F is algebraic over K, then E and F are linearly disjoint over K.

Theorem 9.8. *Let F be an algebraic extension of a field K of characteristic $p \neq 0$ and C an algebraically closed field containing F. Then F is separable over K if and only if F and $K^{1/p}$ are linearly disjoint over K.*

Theorem 9.9. *Let F be an extension field of K of characteristic $p \geq 0$ and C an algebraically closed field containing F. Then the following are equivalent:*

(a) *F and $K^{1/p}$ are linearly disjoint over K;*

(b) *F and K^{1/p^n} are linearly disjoint over K;*

(c) *F and K^{1/p^∞} are linearly disjoint over K;*

(d) *every finitely generated intermediate field E is separably generated over K;*

(e) *$(\mathrm{Aut}_K C)'$ and F are linearly disjoint over K.*

An extension field of F of K is said to be separable *if F satisfies the above equivalent conditions.*

Corollary 9.10. *Let $K \leq E \leq F$ be extension fields.*

(a) *If F is separable over K, then E is separable over K.*

(b) *If F is separable over E and E is separable over K, then F is separable over K.*

(c) *If F is separable over K and E is algebraic over K, then F is separable over E.*

EXERCISE 9.11. Let $K = \mathbb{Z}_p$, $F = \mathbb{Z}_p(x)$, and $E = \mathbb{Z}_p(x^p)$.

(1) F is separably generated and separable over K.

(2) F is algebraic and purely inseparable over E.

(3) $\{x^p\}$ is a transcendence base of F over K which is not a separating transcendence base.

CHAPTER 8

Multivariate polynomials

This chapter is prepared for an expository talk in the 20th Algebra Camp of Seoul National University (June 2008)[1] to introduce the basic notions of *Gröbner bases* with an algorithm for constructing them and to show some applications to several areas of pure and applied mathematics, addressed for beginners without prior knowledge in the subject.

An object to be considered in the Gröbner [2] basis theory is an ideal $I = \langle g_1, \ldots, g_r \rangle$ in the algebra $\mathbb{F}[x_1, x_2, \ldots, x_n]$ of commutative polynomials over a field $\mathbb{F}$, in other words, we deal with polynomial generators in several variables. In Linear Algebra, we study systems of linear equations in several variables, which are solved by *Gaussian elimination.* For example, consider a system of linear equations:

$$2x + 3y + 4z = 5, \quad 3x + 4y + 5z = 2.$$

By using the Gaussian elimination

$$x = z - 14, \quad y = 11 - 2z.$$

In the language of ideal theory, we have

$$\langle 2x + 3y + 4z - 5,\ 3x + 4y + 5z - 2 \rangle = \langle x - z + 14,\ y + 2z - 11 \rangle \subset \mathbb{Q}[x, y, z].$$

On the other hand, in Abstract Algebra, we study (among other things) polynomial equations in one variable. Note that the ring $\mathbb{F}[x]$ of univariate polynomials over a field $\mathbb{F}$ is a principal ideal domain and moreover, a Euclidean domain. For each ideal of $\mathbb{F}[x]$, its generator can be computed by *Euclidean algorithm.* For an ideal of

[1]This was supported by KRF Grant # 2005-070-C00004. The first was initiated in 1999, and it has been held every winter and summer vacation. The author is very grateful for having a chance to learn as a student and a researcher in this joyful camp.

[2]Wolfgang Gröbner(1899–1980) was an Austrian mathematician. His student, Bruno Buchberger(1942–), introduced Gröbner bases in his Ph.D. thesis [B65]. Buchberger named them after his advisor [B76a].

$\mathbb{F}[x_1, x_2, \ldots, x_n]$, we can construct its "good" set of generators by the *Buchberger's algorithm* invented in 1965.

The *division algorithm* for multivariate polynomials should be kept in mind as a motivation to start with the Gröbner basis theory. We present the Gröbner basics[3] and also illustrate that Buchberger's algorithm for computing Gröbner bases provides a fundamental tool for solving a wide range of problems, such as: finding an algebraic relation among polynomials; radical membership problem; the coloring problem of planar graphs.

An historical overview is enumerated for understanding the stream and prospect. Finally, basic references are included, mainly based on Buchberger's work.

1. Monomials

DEFINITION 1.1. (a) A monomial in the polynomial algebra $\mathbb{F}[x_1, x_2, \ldots, x_n]$ over a field $\mathbb{F}$ is of the form $\mathbf{x}^\alpha = x_1^{\alpha_1} x_2^{\alpha_2} \cdots x_n^{\alpha_n}$ ($\alpha_i \in \mathbb{Z}_{\geq 0}$). A monomial with coefficient in $\mathbb{F}^\times$ is called a term.

(b) A monomial ideal of $\mathbb{F}[x_1, x_2, \ldots, x_n]$ is an ideal generated by (possibly infinitely many) monomials.

EXAMPLE 1.2. Let I be a monomial ideal and A be a lattice which consists of exponent vectors belonging to the monomial ideal I. Note that A is additively closed. For an ideal $I = \langle x^2y^3, x^3y \rangle \subset \mathbb{F}[x, y]$, the set A is

$$A = \{(i, j) \in \mathbb{Z}_{\geq 0}^2 \mid (i \geq 2 \text{ and } j \geq 3) \text{ or } (i \geq 3 \text{ and } j \geq 1)\}.$$

Lemma 1.3. *Let $I = \langle \{\mathbf{x}^\alpha | \alpha \in A\} \rangle$ be a monomial ideal. Then*

(a) *$\mathbf{x}^\beta \in I$ if and only if there exists $\alpha \in A$ such that $\mathbf{x}^\alpha | \mathbf{x}^\beta$.*

(b) *$f \in I$ if and only if each term of f lies in I. As corollaries,*

(c) *I is an $\mathbb{F}$-vector space spanned by $\{\mathbf{x}^\alpha \mid \mathbf{x}^\alpha$ is a monomial in $I\}$, and*

(d) *Two monomial ideals are same if and only if they contain the same monomials.*

REMARK 1.4. Let I be a monomial ideal and G be its generating set of monomials. Then G becomes a Gröbner basis for I, which will be defined.

Lemma 1.5 (Dickson's lemma [Dk]). *Any monomial ideal in $\mathbb{F}[x_1, \ldots, x_n]$ is generated by finitely many monomials.* [4]

---

[3]In fact, this is the title of Chapter 1 in [S96].

[4]Another equivalent statement can be found in [AL, Exercise 1.4.12].

Proof. Let $I = \langle\{\mathbf{x}^\alpha | \alpha \in A\}\rangle$ be an ideal. We use induction on m :=the number of variables. First, for the case of $m = 1$, if we take the smallest exponent, call e, then $I = \langle x^e \rangle$.

Let $I \subset \mathbb{F}[x_1, \ldots, x_{m-1}, y]$ be an ideal. The idea of proof is slicing i with respect to y. Let $J \subset \mathbb{F}[x_1, \ldots, x_{m-1}]$ be the ideal generated by the monomials $\mathbf{x}^r$ such that $\mathbf{x}^r y^l \in I$ for some $l \geq 0$. J is the projection of I onto $\mathbb{F}[x_1, \ldots, x_{m-1}]$. By induction hypothesis, J is finitely generated by monomials. Write $J = \langle \mathbf{x}^{\alpha(1)}, \ldots, \mathbf{x}^{\alpha(t)} \rangle$. So $\mathbf{x}^{\alpha(i)} y^{m_i} \in I$ for $1 \leq i \leq t$ and $m := \max\{m_i\}$. We define J_i to be the ideal in $\mathbb{F}[x_1, \ldots, x_{m-1}]$ such that $\mathbf{x}^\beta y^i \in I$ and write $J_i = \langle \mathbf{x}^{\alpha_i(1)}, \ldots, \mathbf{x}^{\alpha_i(t_i)} \rangle$.

We define the following monomial sets:
From the generating set of J_0, we set $G_0 = \{\mathbf{x}^{\alpha_0(1)}, \ldots, \mathbf{x}^{\alpha_0(t_0)}\}$;
from the generating set of J_1, we set $G_1 = \{\mathbf{x}^{\alpha_1(1)} y, \ldots, \mathbf{x}^{\alpha_1(t_1)} y\}$;
$\ldots$;
from the generating set of J_{m-1}, we set $G_{m-1} = \{\mathbf{x}^{\alpha_{m-1}(1)} y^{m-1}, \ldots, \mathbf{x}^{\alpha_{m-1}(t_{m-1})} y^{m-1}\}$;
from the generating set of J, we set $G_m = \{\mathbf{x}^{\alpha(1)} y^m, \ldots, \mathbf{x}^{\alpha(t)} y^m\}$.

Then every monomial in I is divisible by a monomial in $G_0 \cup \cdots \cup G_m$. Thus we have proved. □

Definition 1.6. A *monomial order* (term order) is a total-ordering $<$ on the set of monomials in $\mathbb{F}[x_1, \ldots, x_n]$ such that

(i) $1 < \mathbf{x}^\alpha$ for all monomials $\mathbf{x}^\alpha (\neq 1)$;
(ii) $\mathbf{x}^\alpha < \mathbf{x}^\beta$ implies $\mathbf{x}^\gamma \mathbf{x}^\alpha < \mathbf{x}^\gamma \mathbf{x}^\beta$ for all monomials $\mathbf{x}^\gamma$.

Example 1.7. (1) In a univariate case $\mathbb{F}[x]$, a linear order $<$ satisfying $1 < x < x^2 < \cdots$ is a unique term order.

(2) lexicographic order, degree(graded, homogeneous) lex order, degree reverse lex order, weighted (reverse) lex order:

(i) lexicographic order on $\mathbb{F}[x, y]$ with $x > y$;
$$1 < y < y^2 < \cdots < x < xy < xy^2 < \cdots < x^2 < x^2y < x^2y^2 < \cdots,$$
(ii) degree-lexicographic order on $\mathbb{F}[x, y]$ with $x > y$;
$$1 < y < x < y^2 < xy < x^2 < y^3 < xy^2 < x^2y < x^3 < y^4 < \cdots,$$
(iii) degree reverse lex order on $\mathbb{F}[x, y, z]$ with $x > y > z$;
$$1 < z < y < x < z^2 < yz < xz < y^2 < xy < x^2 < \cdots.$$
Note that $xz > y^2$ with respect to the deg-lex order with $x > y > z$.

REMARK 1.8. Bayer and Stillman proved that the reverse lexicographic order is optimal - as measured by the regularity of the ideal - in computing syzygies [BS87a], and that among the orders which refine a nonstrict ordering (that is, an order that may fail to compare monomials of the same degree) the reverse lexicographic order works best [BS87b].

Lemma 1.9. *Any monomial order in* $\mathbb{F}[x_1, \ldots, x_n]$ *is Artinian, that is, every subset has a least element.*

Proof. ([E95, Lemma 15.2]) Fix a term order $>$. Let A be a nonempty set of monomials. Consider $I := \langle A \rangle$, a monomial ideal. By Dickson's lemma, $I = \langle \mathbf{x}^{\alpha_1}, \ldots, \mathbf{x}^{\alpha_s} \rangle$. Assume $\mathbf{x}^{\alpha_1} < \cdots < \mathbf{x}^{\alpha_s}$. We claim that $\mathbf{x}^{\alpha_1}$ is a minimal element of A. Why?

We choose an arbitrary $m \in A$. Since $m \in I$, m is a multiple of some $\mathbf{x}^{\alpha_i}$. We have $m \geq \mathbf{x}^{\alpha_i} \geq \mathbf{x}^{\alpha_1}$. □

2. Division algorithm

The division algorithm for multivariate polynomials is different from the univariate case. It is slightly complicated. Note that the multivariate polynomial ring $\mathbb{F}[x_1, \ldots, x_n]$ is not a principal ideal domain.

DEFINITION 2.1. Fix a monomial order $<$. Let $f, g \in \mathbb{F}[x_1, \ldots, x_n]$ be given.

(a) The maximal monomial of f with respect to $<$ is called the *leading monomial* of f with respect to $<$, which we denote by $\overline{f} = \mathrm{lm}(f)$. The *leading term* (or *initial term*) of f is denoted by $\mathrm{lt}(f) = \mathrm{in}(f)$. A polynomial f is called monic if the leading coefficient $\mathrm{lc}(f)$ of $\overline{f}$ is 1.

(b) We say that f *reduces* to f' mod g, denoted by $f \underset{g}{\longrightarrow} f'$ if there exists a term t in f such that t is divisible by the leading term $\mathrm{in}(g)$ and $f - \frac{\mathrm{in}(g)}{g} = f'$.

(c) For $G \subset \mathbb{F}[x_1, \ldots, x_n]$, a monomial m is called *G-reduced* or *G-standard* if m is divisible by none of $\mathrm{in}(g)$ for $g \in G$. Otherwise, m is called G-reducible.

EXAMPLE 2.2. $f = xy^2 - x$, $g = y^2 - x$. Fix the deg-lex order $>$ with $x > y$. Then $f - xg = x^2 - x$, i.e., $f \underset{g}{\longrightarrow} x^2 - x$.

Proposition 2.3. *Consider* $>_l, >_{dl}, >_{rl}$*, that is, the lexicographic, degree-lex, degree-reverse-lex monomial orders with* $x_1 > \cdots > x_n$*, respectively. Let* $f \in \mathbb{F}[x_1, \ldots, x_n]$.

(a) *If* $\mathrm{in}_{>_l}(f) \in \mathbb{F}[x_m, \ldots, x_n]$ *for some* m*, then* $f \in \mathbb{F}[x_m, \ldots, x_n]$.

(b) $>_{dl}$ *refines the order by total degree; and if* f *is homogeneous with* $\text{in}_{>_{dl}}(f) \in \mathbb{F}[x_m, \ldots, x_n]$ *for some* m*, then* $f \in \mathbb{F}[x_m, \ldots, x_n]$*.*

(c) $>_{rl}$ *refines the order by total degree; and if* f *is homogeneous with* $\text{in}_{>_{rl}}(f) \in \langle x_m, \ldots, x_n \rangle$ *for some* m*, then* $f \in \langle x_m, \ldots, x_n \rangle$*.*

These properties characterize the orders $>_l, >_{dl}, >_{rl}$ *among monomial orders on* $\mathbb{F}[x_1, \ldots, x_n]$ [E95, §15.2]. *For geometric interpretations of monomial orders, refer to* [E95, §15.8].

Lemma 2.4. *Any order on the set of monomials in* $\mathbb{F}[x_1, \ldots, x_m]$ *is a well ordering if and only if every strictly decreasing sequences of monomials* $\mathbf{x}^{\alpha_1} \geq \mathbf{x}^{\alpha_2} \geq \cdots$ *eventually terminates.*

Proof. ($\Leftarrow$) Suppose not, i.e., there exists a nonempty set S of monomials without least element. Pick $\mathbf{x}^{\alpha_1} \in S$. Then $\mathbf{x}^{\alpha_1}$ is not the smallest. $\mathbf{x}^{\alpha_1} > \mathbf{x}^{\alpha_2}$. Continue to have a contradiction.

($\Rightarrow$) Suppose not, i.e., there exists an infinite sequence $\mathbf{x}^{\alpha_1} > \mathbf{x}^{\alpha_2} > \cdots$. Then $\{\mathbf{x}^{\alpha_1}, \mathbf{x}^{\alpha_2}, \ldots\}$ has no minimal element. □

Theorem 2.5 (Division algorithm). *Let* $G \subset \mathbb{F}[x_1, \ldots, x_n]$*. Fix a monomial order. Then every* $f \in \mathbb{F}[x_1, \ldots, x_n]$ *can be written as*

$$f = \sum \alpha_i a_i g_i + \sum \beta_j u_j, \tag{2.1}$$

where $\alpha_i, \beta_j \in \mathbb{F},\ a_i, u_j,$ *monomials,* $g_i \in G,\ a_i\overline{g_i} \leq \overline{f}, u_j \leq \overline{f}, u_j$: *G-standard.*

Definition 2.6. In (2.1), $r = \sum \beta_j u_j$ is called a *normal form* (or a remainder) of f with respect to G, denoted by $\text{NF}_G(f)$ if it is unique or when we impose the order of divisions by elements in G.

Remark 2.7. (1) By the well-ordering property of monomial orders, the division algorithm terminates in a *finite* number of steps.

(2) In general, a normal form of f is *not* unique. A normal form in the multivariate polynomial division depends on the order of divisions. For instance, divide $xy^2 - x$ by $xy - y$ and $y^2 - x$. On the other hand, divide $xy^2 - x$ first by $y^2 - x$ and then by $xy - y$. Compare the results. Why does this occur? How can you solve this problem?

Corollary 2.8. *With the same assumptions as Theorem 2.5, the set of all G-standard monomials in* $\mathbb{F}[x_1, \ldots, x_n]$ *spans the* $\mathbb{F}$*-vector space* $\mathbb{F}[x_1, \ldots, x_n]/\langle G \rangle$*.*

3. Gröbner bases

Problems. Let A be a commutative algebra over a field $\mathbb{F}$ and I be an ideal of A generated by $g_1, \dots, g_r$.

(1) Given $f \in A$, determine whether $f \in I$ or not. (Ideal membership problem)

(2) Find an $\mathbb{F}$-linear basis of A/I and its multiplication table, i.e., determine the algebra structure of A/I. For example, we consider

$$I = \langle xy - y, y^2 - x \rangle \subset \mathbb{Q}[x, y].$$

DEFINITION 3.1. Let I be an ideal of $\mathbb{F}[x_1, \dots, x_n]$. The *initial ideal* $\mathrm{in}_>(I)$ is a monomial ideal generated by the initial terms $\mathrm{in}_>(f)$ (or the leading terms $\mathrm{lt}_>(f)$) for all $f \in I$.

EXAMPLE 3.2. (1) Let $I = \langle xy+1, x^2 \rangle \subset \mathbb{F}[x, y]$ and fix lex order $>$ with $x > y$. We claim that $\langle xy, x^2 \rangle \subsetneq \mathrm{in}(I)$. This is because $x, 1 \in \mathrm{in}(I) \setminus \langle xy, x^2 \rangle$.

(2) Let $I = \langle x+z, y-z \rangle \subset \mathbb{F}[x, y, z]$ and fix (graded) lex order $>$ with $x > y > z$. We claim that $\mathrm{in}(I) = \langle x, y \rangle$.

Theorem 3.3 (Macaulay[5]). *The set of all monomials not in* $\mathrm{in}_>(I)$ *is an* $\mathbb{F}$*-basis for* $\mathbb{F}[x_1, \dots, x_n]/I$.

Proof. See [E95, Theorem 15.3]. □

Lemma 3.4. *If* $J \subset I \subset \mathbb{F}[x_1, \dots, x_n]$ *are ideals and* $\mathrm{in}_>(J) = \mathrm{in}_>(I)$ *with a monomial order* $>$*, then* $J = I$*. In other words, any set of polynomials in* $\mathbb{F}[x_1, \dots, x_n]$ *whose initial terms generate* $\mathrm{in}_>(I)$ *actually generate* I.

Proof. Suppose that $J \subsetneq I$. Then we choose a polynomial $f \in I \setminus J$ whose initial term is smallest among initial terms of all polynomials in $I \setminus J$. Since $\mathrm{in}(f) \in \mathrm{in}(I) = \mathrm{in}(J)$, $\mathrm{in}(f) = \mathrm{in}(g)$ for some $g \in J$. But $f - g \in I$ and $f - g \notin J$. So $f - g$ has smaller initial terms than f, which is a contradiction. □

REMARK 3.5. Given an ideal $I \subset \mathbb{F}[x_1, \dots, x_n]$ and an arbitrary polynomial $f \in \mathbb{F}[x_1, \dots, x_n]$, once we have a method to compute initial ideals, then we can decide whether $f \in I$ by Lemma 3.4 because $f \in I$ if and only if $\mathrm{in}(I) = \mathrm{in}\,(I + \langle f \rangle)$.

[5]Francis Sowerby Macaulay(1862–1937), an English mathematician who made significant contributions to Algebraic Geometry.

DEFINITION 3.6. Let $g_1, \dots, g_r \in \mathbb{F}[x_1, \dots, x_n]$ and consider $I = \langle g_1, \dots, g_r \rangle$. Fix a monomial order $>$. Then we say that the set $\{g_1, \dots, g_r\}$ is a *Gröbner basis* for I if the following equivalent conditions are satisfied:

(a) $\mathrm{in}_>(I) = \langle \mathrm{in}(g_1), \dots, \mathrm{in}(g_r) \rangle$;

(b) standard monomials w.r.t. $g_1, \dots, g_r$ form an $\mathbb{F}$-basis of $\mathbb{F}[x_1, \dots, x_n]/I$;

(c) for each $f \in I$, the normal form $\mathrm{NF}(f)$ of f w.r.t. $g_1, \dots, g_r$ is 0.

REMARK 3.7. (1) If $G := \{g_1, \dots, g_r\}$ is a Gröbner basis, then every element in $\mathbb{F}[\mathbf{x}]/\langle G \rangle$ can be expressed uniquely in an $\mathbb{F}$-linear combination of G-standard monomials. The characterization (c) of Gröbner bases appears in [B76a]. It is not difficult to show that the above three conditions (a),(b),(c) are equivalent.

(2) Note that the initial ideal $\mathrm{in}_>(I)$ depends on the choice of monomial order $>$. See [BMo] and [MoR] for characterizations of the finite set of $\mathrm{in}_>(I)$ as the order $>$ varies, where the notions of state polytope and Gröbner fan are defined.[6] For universal Gröbner bases and some of the results involving the variation of monomial orders, see also the text [S96, Ch.2–3].

Theorem 3.8 (Existence of Gröbner bases). *Every ideal $I \subset \mathbb{F}[\mathbf{x}]$ has a Gröbner basis.*

Proof. Apply Dickson's lemma to $\mathrm{in}_>(I)$. Then there exist monomials $m_1, \dots, m_t \in \mathrm{in}(I)$ generating $\mathrm{in}(I)$. So there exist $g_1, \dots, g_t \in I$ such that $\mathrm{in}(g_1) = m_1, \dots, \mathrm{in}(g_t) = m_t$. We claim that $\{g_1, \dots, g_t\}$ is a generating set for I.

Let $f \in I$. Divide f by $g_1, \dots, g_t$ to have $f = q_1 g_1 + \cdots + q_t g_t + r$. Suppose $r \neq 0$. Then $\mathrm{in}(r) \in \mathrm{in}(I)$ and $\mathrm{in}(g_i) | \mathrm{in}(r)$ for some i. This contradiction proves our claim. □

Corollary 3.9 (Hilbert's basis theorem). [7] *Every ideal of $\mathbb{F}[x_1, \dots, x_n]$ is finitely generated.*

Corollary 3.10. *$\mathbb{F}[\mathbf{x}]$ is a Noetherian ring, i.e., there is no infinite ascending chain of ideals.*

---

[6] The tropical variety $T(I)$ of an ideal I is a subfan of the Gröbner fan of I. [BJSST]

[7] The proof using the fact that every ideal of $\mathbb{F}[\mathbf{x}]$ has a finite Gröbner basis is first due to P. Gordan in 1899. For the controversy between P. Gordan and D. Hilbert on the value of existence proofs, one may refer to Ch.5 in [C. Reid, *Hilbert* (이일해 역, 대우학술총서·번역 **19**, 민음사, 1989)]. Also, an educational book [박부성 저, *천재들의 수학노트: 괴짜 천재들의 수학과 인생*, 향연, 2003] contains one thing or another interesting accounts on David Hilbert or other great mathematicians.

Proof. Suppose that there is a chain of ideals $I_0 \subsetneq I_1 \subsetneq I_2 \subsetneq \cdots$. Let $I = \bigcup_{i\geq 0} I_i$, ideal, finitely generated by the above Corollary. So $I = \langle h_1, h_2, \ldots, h_s \rangle \subset \bigcup_{i=0}^{l} I_i$ for some l. We have $I = \bigcup_{i=0}^{l} I_i$, which is a contradiction. □

DEFINITION 3.11. A Gröbner basis G of I is called *minimal* if $\text{in}(g)$ does not divide $\text{in}(g')$ for all distinct $g, g' \in G$, and G is called *reduced* if no term in g' is divisible by any $\text{in}(g)$ for all distinct $g, g' \in G$.

Theorem 3.12 (Buchberger). *Fix a monomial order. Then every ideal has a unique reduced Gröbner basis.*

Proof. See [AL, §1.8] or [B76b]. □

Our natural question is how to construct Gröbner bases. Let an ideal $I = \langle G \rangle$ be given. The basic idea is: to enrich the leading term structure, we add another polynomial in I to a given generating set G to have an enlarged set of generators. But an obvious trouble is that this process involves infinite testing. Buchberger's ingenuity is: we have only to check a *finite* number of polynomials in I, leading us an algorithm for constructing Gröbner bases.

DEFINITION 3.13. The S-polynomial of monic polynomials g_i and g_j is defined by

$$S(g_i, g_j) := m_{ji} g_i - m_{ij} g_j, \quad \text{where} \quad m_{ij} = \text{in}(g_i)/\gcd(\text{in}(g_i), \text{in}(g_j)).$$

Note that the S-polynomial of g_i and g_j is the smallest polynomial combination of g_i and g_j canceling the leading terms of g_i and g_j.

Theorem 3.14 (Buchberger 1965). *index*[8] *Let*

$$S(g_i, g_j) = \sum_{u=1}^{r} f_u^{(ij)} g_u + h_{ij} \quad \Big(\deg(h_{ij}) < \deg(f_u^{(ij)} g_u) \leq \deg(m_{ji} g_i) \Big)$$

be a standard expression by division algorithm. Then the set $\{g_1, \ldots, g_r\}$ *is a* Gröbner basis *for* I *if and only if* $h_{ij} = 0$ *for all* $i \neq j$.

Proof. See [AL, Theorem 1.7.4] or [E95, Theorem 15.8]. □

[8] Recently, in honor of Bruno Buchberger (60th birthday), an English translation of his Ph.D. thesis (*An algorithm for finding the basis elements of the residue class ring of a zero dimensional polynomial ideal*) appeared in J. Symbolic Comput. **41** (2006) 475–511.

Corollary 3.15 (Buchberger's algorithm). *Buchberger's theorem gives an algorithm for finding a set of generators $g_1, \ldots, g_r$ for I such that standard monomials form an $\mathbb{F}$-basis of $\mathbb{F}[x_1, \ldots, x_n]/I$ as follows :*

INPUT: $F = \{f_1, \ldots, f_r\}$
OUTPUT: $G = \{g_1, \ldots, g_s\}$, a Gröbner basis

$G := F$
Repeat
    $G' = G$
    For each pair $\{p, q\}$, $p \neq q$ in G'. DO
        $s' := \mathrm{NF}_G(S(p, q))$
        If $s' \neq 0$, then $G := G \cup \{s'\}$.
Until $G = G'$.

REMARK 3.16. For a finite set of generator for an ideal, we can compute its Gröbner basis in a *finite* number of steps by Dickson's lemma.

EXAMPLE 3.17. Let $I = \langle xy - y, y^2 - x\rangle \subset \mathbb{Q}[x, y]$. Find a $\mathbb{Q}$-linear basis of $\mathbb{Q}[x, y]/I$ and its multiplication table. The solution is as follows:

Let $f_1 = xy - y$, $f_2 = y^2 - x$ and $F = \{f_1, f_2\}$. Fix the degree-lex order with $x > y$. $S(f_1, f_2) = yf_1 - xf_2 = x^2 - y^2 \xrightarrow{F} x^2 - x =: f_3$. Add f_3 to F and let $F' = \{f_1, f_2, f_3\}$. Then $S(f_1, f_2) \xrightarrow{F'} 0$, $S(f_1, f_3) = 0$, and $S(f_2, f_3) = -x^3 + xy^2 \xrightarrow{F'} 0$. Thus $F' = \{xy - y, y^2 - x, x^2 - x\}$ is a Gröbner basis.

The set of I-standard monomials is $\{1, x, y\}$, which is a $\mathbb{Q}$-linear basis of $\mathbb{Q}[x, y]/I$.

The multiplication table is

	1	x	y
1	1	x	y
x		x	y
y			x

. □

REMARK 3.18. The computer algebra system SINGULAR provides the commands `> std(I);` or `> groebner(I);` for computing the reduced Gröbner basis of an ideal I. In the following example, we obtain the reduced Gröbner basis of $I = \langle yz^2, y^2 + xz\rangle \subset \mathbb{Q}[x, y, z]$ with respect to degree reverse lexicographic order $x > y > z$:

```
> ring R = 0, (x,y,z), dp;
> ideal I = yz2, y2+xz;
> std(I);
```

```
_[1]=y2+xz
_[2]=yz2
_[3]=xz3 .
```

If one wants to use the degree lexicographic order or purely lexicographic order then one has only to replace `dp` by `Dp` or `lp`, respectively.

EXERCISE 3.19. Consider $F := \{f_1, f_2\} \subset \mathbb{Q}[x, y]$, where

$$f_1 = x^2y - y + x, \quad f_2 = xy^2 - 2x.$$

Determine if $p := -4xy + xy^3 + 2y^2 - 4x^2$ belongs to the ideal $\langle f_1, f_2\rangle$. If it does, then find $q_1, q_2 \in \mathbb{Q}[x, y]$ such that $p = f_1q_1 + f_2q_2$.

EXERCISE 3.20. Use SINGULAR to compute the reduced Gröbner basis of $\{f_1, f_2, f_3\}$ w.r.t. lex order with $x > y > z$, where

$$\begin{aligned} f_1 &= x^2 + y^2 + z^2 - 1, \\ f_2 &= x^2 + y^2 + z^2 - 2x, \\ f_3 &= 2x - 3y - z. \end{aligned}$$

Use this result to solve the system of nonlinear equations:

$$x^2 + y^2 + z^2 = 1, \; x^2 + y^2 + z^2 - 2x = 0, \; 2x - 3y - z = 0.$$

REMARK 3.21. (1) We can save the computation time in the basic Buchberger's algorithm by detecting trivial reductions of S-polynomials. This improved algorithm, known as Buchberger's *normal selection strategy*, is developed in [B79]. Also, Gebauer and Möller describe practical criteria for detecting superfluous S-polynomial reductions [GM]. On the other hand, Giovini et al. suggest an improved version of Buchberger's algorithm, called the *sugar strategy*, that seeks to exploit the strategic benefits of homogenization [GMNRT].

(2) The complexity for computing Gröbner bases is doubly exponential with respect to the number of variables. A computational issue of Buchberger's algorithm appears, for example, in [BM, BS87a, BS88, B83, MaMe, MM, W].

(3) Buchberger and Möller gave an algorithm for computing a Gröbner basis and a quotient basis of the ideal of all polynomials which vanish at a given finite set of points in k^n (k, a field) [BMö].

(4) The lexicographical Gröbner basis can be obtained by applying the FGLM algorithm after a total degree Gröbner basis computation [FGLM]. It is usually much

faster to compute the basis this way than with a direct application of Buchberger's algorithm. Also refer to the methods of Gröbner walk [CKM] and Hilbert driven [Tr].

(5) If we consider noncommutative polynomial algebras, then there is no guarantee that the Buchberger algorithm works well, which means that the process may not terminate in finite steps [Be, GMU, Mo, U]. However, for a class of *solvable* noncommutative polynomial rings, an extension of Buchberger's algorithm that computes Gröner bases of one- and two-sided ideals does work [K-RW]. Computational methods for commutative algebras can be applied to noncommutative algebras with PBW bases such as iterated Ore extensions of a division ring[9] and enveloping algebras of finite dimensional Lie algebras [10]. For a detailed exposition on PBW algebras and their computational topics, one may consult the lecture note by H. Li [Li] and the dissertation by V. Levandovskyy [Le].

4. Some applications

In this section, we present some applications of Gröbner basis theory to various areas of mathematics and related fields.

4.1. Integer programming.

Consider the following problem: Minimize the linear function $P + N + D + Q$ subject to $P, N, D, Q > 0$ integers and

$$P + 5N + 10D + 25Q = 117.$$

This problem has a unique solution

$$(P, N, D, Q) = (2, 1, 1, 4).$$

Let me explain more precisely.

- Represent a collection C of coins by a monomial $p^a n^b d^c q^d$ in the variables p, n, d, q. E.g., 2 pennies and 4 dimes is $p^2 d^4$.

- Input set $F = \{p^5 - n, p^{10} - d, p^{25} - q\}$. Represents the basic relationships among coins.

---

[9] Skew polynomial rings and Weyl algebras belong to this class.

[10] A standard textbook [H72, §17] contains a nontrivial proof on the existence of PBW bases in universal enveloping algebras of finite dimensional Lie algebras.

- Output set $G = \{p^5 - n, n^2 - d, nd^2 - q, d^3 - nq\}$. Expresses a more useful set of replacement rules. E.g., the expression $d^3 - nq$ translates to: replace 3 dimes with a nickel and a quarter.

- Given a collection C of coins, we use rules encoded by G to transform (in any order) C into a set of coins C' with equal monetary value but smaller number of elements.

EXAMPLE 4.1 (Solving the previous integer program).

$$\begin{array}{ccccccc} p^{17}n^{10}d^5 & \longrightarrow & p^{12}n^{11}d^5 & \longrightarrow & \cdots & \longrightarrow & p^2n^{13}d^5 \\ & & & & & & \downarrow \\ p^2ndq^4 & \longleftarrow & \cdots & \longleftarrow & p^2n^{11}dq^2 & \longleftarrow & p^2n^{12}d^3q \end{array}$$

REMARK 4.2. Gröbner bases give a method of transforming a feasible solution using local moves into a global optimum. E.g.,

$$\{p^5 - n, p^{10} - d, p^{25} - q\} \underset{algorithm}{\overset{Buchberger}{\longrightarrow}} \{p^5 - n, n^2 - d, nd^2 - q, d^3 - nq\}, \quad \text{a Gröbner basis.}$$

4.2. Invariant theory [S93, Ch.1].

A problem in invariant theory is: What are elements in $\mathbb{C}[x_1, \ldots, x_n]$ which are invariant under the action of a finite group $\Gamma \subset \mathrm{GL}_n(\mathbb{C})$? It is well known that the invariant ring $\mathbb{C}[\mathbf{x}]^\Gamma$ is generated by a finite set of fundamental invariants [S93, Theorem 2.1.3].

Question. Compute all algebraic relations between the fundamental invariants for the invariant ring of a finite group $\Gamma \subset \mathrm{GL}_n(\mathbb{C})$.

EXAMPLE 4.3. A polynomial in $\mathbb{C}[x_1, \ldots, x_n]$ is said to be symmetric if it is invariant under every permutation of the variables $x_1, \ldots, x_n$. The symmetric polynomials form a subring $\mathbb{C}[\mathbf{x}]^{S_n}$ of $\mathbb{C}[\mathbf{x}]$. Consider the symmetric polynomials in 3 variables, among which,

$$x_1 + x_2 + x_3, \ x_1x_2 + x_2x_3 + x_3x_1, \ x_1x_2x_3$$

are called elementary symmetric polynomials. Are these three polynomials algebraically independent?

We apply the slack variable computation. Using the lex order with $e_1 < e_2 < e_3 < x_1 < x_2 < x_3$, we calculate from

$$G = \{x_1 + x_2 + x_3 - e_1,\ x_1x_2 + x_2x_3 + x_3x_1 - e_2,\ x_1x_2x_3 - e_3\},$$

to obtain its Gröbner basis

$$\mathcal{G} = \{x_3 + x_2 + x_1 - e_1,\ x_2^2 + x_1x_2 + x_1^2 - e_1x_2 - e_1x_1 + e_2,\ x_1^3 - e_1x_1^2 + e_2x_1 - e_3\}.$$

This does not contain any polynomial only in the slack variables e_1, e_2, e_3. Thus, the elementary symmetric polynomials have no algebraic relation among e_1, e_2, e_3, i.e., they are algebraically independent. In fact, $\mathbb{C}[\mathbf{x}]^{S_n} = \mathbb{C}[e_1, \ldots, e_n]$, which is called the *fundamental theorem of symmetric polynomials.*

We also use the Gröbner basis normal form reduction to obtain that $(x_1-x_2)^2(x_1-x_3)^2(x_2-x_3)^2$ reduces to $-27e_3^2 + 18e_3e_2e_1 - 4e_3e_1^3 - 4e_2^3 + e_2^2e_1^2$.

Example 4.4. Let us consider $\mathbb{Z}_4 = \left\{\pm\begin{pmatrix} 1 & 0 \\ 0 & 1 \end{pmatrix}, \pm\begin{pmatrix} 0 & -1 \\ 1 & 0 \end{pmatrix}\right\}$. Then its invariant ring equals

$$\mathbb{C}[x_1, x_2]^{\mathbb{Z}_4} = \{f \in \mathbb{C}[x_1, x_2] \mid f(x_1, x_2) = f(-x_2, x_1)\}.$$

We have a set of fundamental invariants

$$\{I_1 := x_1^2 + x_2^2,\ I_2 := x_1^2x_2^2,\ I_3 := x_1^3x_2 - x_1x_2^3\}.$$

Using the slack variable Gröbner basis method, we get an algebraic relation $I_3^2 + 4I_2^2 - I_2I_1^2 = 0$. In fact, $\mathbb{C}[x_1, x_2]^{\mathbb{Z}_4} = \mathbb{C}[I_1, I_2] \oplus I_3\mathbb{C}[I_1, I_2]$.

4.3. Radical membership problem [CLO07, §4.2].

Problem. Let $f \in k[x_1, \ldots, x_n]$ and I be an ideal of $k[x_1, \ldots, x_n]$. If k is an algebraically closed field, does f vanish identically on $V(I)$, i.e., $f \in I(V(I)) = \sqrt{I}$?

Theorem 4.5 (Weak Nullstellensatz). *Let I be an ideal of $k[x_1, \ldots, x_n]$. Then*

$$V_{\bar{k}}(I) = \emptyset \text{ if and only if } I = k[x_1, \ldots, x_n].$$

Example 4.6. (1) Can 1 be written as a combination of $1 - xy$ and y^2 in $k[x, y]$?

(2) Let $f = 1 - xy$ and $g = x - x^2y + y^2 - xy^3 + x^2$. Do f and g have a common zero over $\mathbb{C}$?

Definition 4.7. Let I be an ideal of a ring R. Then the radical of I is defined to be

$$\sqrt{I} := \{x \in R| \ x^n = 0 \text{ for some } n > 0\},$$

which is an ideal of R. We call I a radical ideal if $I = \sqrt{I}$.

Theorem 4.8 (Strong Nullstellensatz). *Let I be an ideal of $k[x_1, \ldots, x_n]$. Then*

$$I(V_{\overline{k}}(I)) = \sqrt{I}.$$

Remark 4.9. Consider $V(x(x-t), y)$ $(t \in k)$. Then $I_t = \langle x(x-t), y\rangle$ is a radical ideal unless $t = 0$. We deform by taking $t \to 0$ to obtain $V(x^2, y)$ and $I_0 = \langle x^2, y\rangle$ $(\sqrt{I_0} = \langle x, y\rangle)$. A double point is not distinguishable with a single point from the viewpoint of geometry, but should be regarded as two points from an algebraic viewpoint, motivating the scheme structure.

Theorem 4.10. *Let $I = \langle f_1, \ldots, f_m\rangle$ be an ideal of $k[x_1, \ldots, x_n]$ and $f \in k[x_1, \ldots, x_n]$. Then $f \in \sqrt{I}$ if and only if $1 \in \langle f_1, \ldots, f_m, 1 - yf\rangle \subseteq k[x_1, \ldots, x_n, y]$.*

Proof. [11] First, let $f \in \sqrt{I} = I(V_{\overline{k}}(I))$. We show that the polynomials $f_1, \ldots, f_m, 1-yf$ have no common zeros in $\overline{k}^{n+1}$ by Weak Nullstellensatz. Suppose not, i.e., there exists $(a_1, \ldots, a_n, b)$ such that $f_1(a_1, \ldots, a_n) = 0, \cdots, f_m(a_1, \ldots, a_n) = 0$ and $f(a_1, \ldots, a_n) = 1/b \neq 0$. This contradicts the fact that $f \in I(V_{\overline{k}}(f_1, \ldots, f_m))$, then we obtain that $\langle f_1, \ldots, f_m, 1 - yf\rangle = k[x_1, \ldots, x_n, y]$.

For the converse, $1 \in \sum_{i=1}^m h_i f_i + h(1 - yf)$ for some $h_i, h \in k[x_1, \ldots, x_n, y]$. We show that $f(a_1, \ldots, a_n) = 0$ for all $(a_1, \ldots, a_n) \in V_{\overline{k}}(I)$. For a point $(a_1, \ldots, a_n) \in V_{\overline{k}}(I)$, we calculate at $(a_1, \ldots, a_n, y)$ to have a polynomial in y, $1 = h(a_1, \ldots, a_n, y)(1-yf(a_1, \ldots, a_n))$. So we must have $f(a_1, \ldots, a_n) = 0$. □

Remark 4.11. This theorem gives an algorithm to determine if $f \in \sqrt{I}$:

Input: $f_1, \ldots, f_m, f \in k[x_1, \ldots, x_n]$

Output: $\begin{cases} \text{Yes, if } f \in \sqrt{\langle f_1, \ldots, f_m\rangle}, \\ \text{No, if } f \notin \sqrt{\langle f_1, \ldots, f_m\rangle}. \end{cases}$

(1) Compute the Gröbner basis G of $\langle f_1, \ldots, f_m, 1 - yf\rangle \subseteq k[x_1, \ldots, x_n, y]$.

(2) Yes if $G = \{1\}$, and No otherwise.

---

[11] This is the "trick of Rabinowitch". See [CLO07, §4.2 Theorem 2].

EXAMPLE 4.12. Let $I = \langle x^4y^2 + z^2 - 4xy^3z - 2y^5z, x^2 + 2xy^2 + y^4 \rangle \subset k[x, y, z]$ and $f = yz - x^3$. Does f vanish identically on $V(I)$? Find the least power of f which lies in I.

Solution. To check if $f \in I(V(I)) = \sqrt{I}$, we determine if $1 \in \langle I, 1-wf \rangle \subseteq k[x, y, z, w]$ by using SINGULAR.

```
> ring r=0, (x,y,z,w), dp;
> poly f1=x4y2+z2-4xy3z-2y5z;
> poly f2=x2+2xy2+y4;
> poly f=yz-x3;
> ideal J=f1,f2,1-w*f;
> std(J);   _[1]=1
```
which means that $f \in \sqrt{I}$.

```
> ideal I=f1,f2;
> NF(f,std(I));   x3-yz
> NF(f^2,std(I));   2x2y3z+2x3yz+2y2z2+2xz2
> NF(f^3,std(I));   0
```
which means that $f^3 \in I$.

Directly, we can use the built-in command to test the radical containment:

```
> LIB "poly.lib";
> rad_con(f,I);   1
```
The output is 1 iff $f \in \sqrt{I}$. □

4.4. The 3-color problem [AL, §2.7]. [12]

Problem. Determine whether a given graph can be 3-colored. That is, for a graph with vertices and with at most one edge between any two vertices, can we put the vertices with 3 colors such that no adjacent vertices are of the same color?

For example, we consider the following graph.

[12] This application appeared first in D. Bayer's thesis [Ba, Ch.III, Example 2.3]. For a general result on graph colorability, refer to the paper [HW].

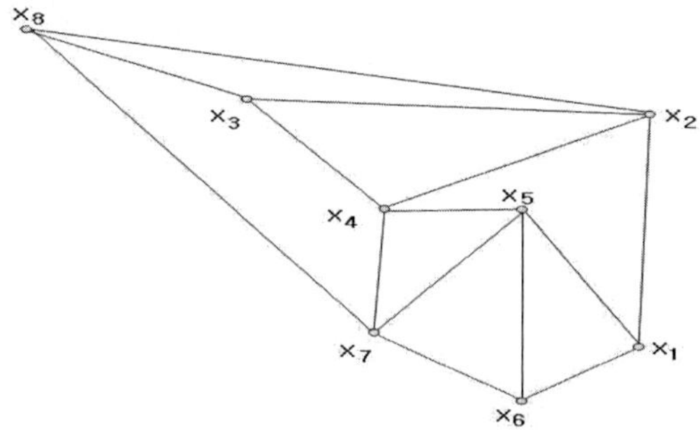

Each vertex x_i $(1 \leq i \leq 8)$ is to be assigned one of the 3 colors. We represent the 3 colors by the 3 distinct cube roots of unity, i.e., we have equations $x_i^3 - 1 = 0$ $(1 \leq i \leq 8)$. If the vertices x_i and x_j are connected by an edge, then they need to have in a different color. That means, $x_i^2 + x_i x_j + x_j^2 = 0$ for $i \neq j$ $(\because x_i^3 = x_j^3 \Rightarrow (x_i - x_j)(x_i^2 + x_i x_j + x_j^2) = 0)$. So the corresponding set of polynomials in $\mathbb{C}[x_1, \ldots, x_8]$ is

$$G = \{x_i^3 - 1,\ x_r^2 + x_r x_s + x_s^2 \mid 1 \leq i \leq 8,\ (r, s) = (1,2), (1,5), (1,6), (2,3), (2,4), (2,8), (3,4), (3,8), (4,5), (4,7), (5,6), (5,7), (6,7), (7,8)\}.$$

Theorem 4.13. *The graph is 3-colorable if and only if the zero set $V(G) \neq \emptyset$. By Hilbert's Nullstellensatz, $V(G) \neq \emptyset$ if and only $\langle G \rangle$ is a proper ideal of $\mathbb{C}[\mathbf{x}]$, i.e., $1 \notin \langle G \rangle$.*

In our case, fixing the deg-lex order with $x_1 > x_2 > \cdots > x_8$, we obtain the reduced Gröbner basis

$$\mathcal{G} = \{x_1 - x_7, x_2 + x_7 + x_8, x_3 - x_7, x_4 - x_8, x_5 + x_7 + x_8, x_6 - x_8, x_7^2 + x_7 x_8 + x_8^2, x_8^3 - 1\}.$$

Since $1 \notin \langle G \rangle$, we conclude that this graph is 3-colorable.

REMARK 4.14. It is well known that every planar graph is 4-colorable, which is proved by K. Appel and W. Haken in 1977.

4.5. Multidimensional systems.

In [B85, B01b], B. Buchberger presents an easy introduction to the theory of Gröbner bases and its applications to multidimensional systems theory by giving concrete examples. The seminal paper [O] shows how exactly some fundamental problems of systems theory can be reduced to the problem of constructing Gröbner bases.

4.6. Geometric engineering.

The paper [B88] contains several interesting applications in nonlinear computational geometry, for example, inverse kinematics in robot programming, collision detection for superellipsoids, detection of singularity, etc. Related materials on robotics and geometry are described in the book [CLO07, Ch.6] at an elementary level.

For some papers on geometry theorem proving, one may refer to *Algebraic Approaches to Geometric Reasoning* (H. Hong, D. Wang and F. Winkler eds.), Ann. Math. Artificial Intelligence **13** (1995), Special issue on the workshop (Schloss Weinberg, 1992), 1–202.

Exercises

1. Let A_n be the nth *Weyl algebra*, i.e., the noncommutative $\mathbb{C}$-algebra generated by $x_1, \dots, x_n$ and $\partial_1, \dots, \partial_n$, where all pairs of variables commute, except $\partial_i x_i = 1 + x_i \partial_i$ for each $i = 1, \dots, n$. Find a definition of a Gröbner basis of a 2-sided ideal in A_n. How must one modify Buchberger's algorithm to compute such a Gröbner basis? [13]
2. Let $\mathfrak{E}_n$ be the k-algebra (k, a field) generated by elements $x_1, \dots, x_n$ satisfying the skew-symmetric relations: $x_i x_j = -x_j x_i$ if $i \neq j$, and $x_i^2 = 0$ for each $i = 1, \dots, n$. This finite dimensional k-algebra $\mathfrak{E}_n$ is called the *exterior algebra* with n variables.
   (1) Define a notion of Gröbner basis in $\mathfrak{E}_n$ and give an algorithm to compute it.
   (2) Take a simplicial complex Δ and let $I_\Delta \subset \mathfrak{E}_n$ be the *face ideal*, i.e., the ideal generated by the monomials whose support are the non-faces in Δ, which is called also *Stanley-Reisner ideal* (See [Sta, Ch.2]). Find a primary decomposition of I_Δ [Sch, Theorem 5.3.3].
   (3) * Note that I_Δ is a monomial ideal in $\mathfrak{E}_n$. Now make a random linear change of coordinates. The initial ideal of this new ideal is the face ring of a simplicial complex. What is it?

[13] Some basics on Gröbner bases of modules over the Weyl algebra are reviewed in [Wa, §3].

Historical overview

The following[14] is an enumeration of brief history on the notion of Gröbner bases.

- Euclid: Non-linear univariate systems. Euclid's algorithm turns out to be a special case of the Gröbner bases algorithm.
- C. F. Gauss: Linear multivariate systems. Gauss's algorithm turns out to be a special case of the Gröbner bases algorithm.
- Division Process in 19th Century: A kind of division process for multivariate polynomials.
- P. Gordan 1899: Introduced (in the language of that time) the notion of Gröbner bases (based on the division process) and showed their existence and finiteness, but did not give an algorithm for constructing them.
- L. E. Dickson 1913: A general lemma on the termination of certain sequences of tuples of natural numbers. From this lemma, the termination of the Gröbner bases algorithm follows easily. And, then, Hilbert's basis theorem is an easy corollary of the existence of Gröbner bases.
- M. Janet 1924: Similar notions and methods for differential equations (viewed symbolically).
- G. Hermann 1926: A solution of the membership problem for polynomial ideals based on linear algebra.
- W. Gröbner 1954: The method in his seminar 1964 was already published in one of his papers (1954): "I am using this method already for 17 years..."
- A. I. Shirshov 1962: Similar notion and method for Lie algebras, so called, the Composition Lemma.
- H. Hironaka 1964: In the frame of his famous paper on the resolution of singularities, the notion of "standard bases" in the domain of formal power series, but no algorithm for their construction.
- B. Buchberger 1965: Introduce the notion of S-polynomials and present an algorithm for finding Gröbner bases.
- L. Bokut 1976: Apply Shirshov's method to associative algebras.
- G. M. Bergman 1978: Used the Diamond Lemma for ring theory.
- W. T. Wu 1978: Ritt characteristic sets.

[14] A part of the section is extracted from `http://www.ricam.oeaw.ac.at/Groebner-Bases-Bibliography/`.

- B. Buchberger, H. M. Möller 1982: Construct an algorithm for finding Gröbner bases of multivariate polynomials with preassigned zeros.
- J.-C. Faugère, P. Gianni, D. Lazard, T. Mora 1993: Efficient computation of zero-dimensional Gröbner bases by change of ordering.
- J.-C. Faugère 1999, 2002: New efficient algorithms for computing Gröbner bases (F_4, F_5).
- S.-J. Kang, K.-H. Lee 2000: Introduce the notion of Gröbner-Shirshov pairs for the representations of associative algebras.
- S.-J. Kang, D.-I. Lee, K.-H. Lee, H. Park 2007: Construct linear algebraic algorithms for finding Gröbner-Shirshov bases for noncommutative algebras and their representations.
- S.-J. Kang, D.-I. Lee, E. Park, H. Park 2010: Define a generalized notion of Castelnuovo-Mumford regularity for representations of noncommutative algebras.
- .... others.

A reference guide

It is highly recommended to refer to overview articles on Gröbner bases written by B. Buchberger [B98, B01a], which lead one to have a feeling why and how the Gröbner basis theory is used so widely and essentially in mathematical problems. The conference proceeding [BW] written by leading experts provides fourteen tutorial lectures and seventeen research papers relating to applications in various areas of mathematics. For original developments of the theory, one may directly consult recent translations of Buchberger's Ph.D. thesis and its journal publication [B65, B70]. The papers [B87, BLRW] would be helpful for an overview to polynomial ideal theory.

An excellent survey paper [BM] written by Bayer and Mumford provides an introduction to some of the main computational problems, for which the first two sections deal with Gröbner bases from a geometric setting and from an algebraic point of view. The articles [DS, Gl] will present nice survey on a large part of Computer Algebra and Algebraic Geometry with an emphasis on Gröbner bases.

Introductory materials on Gröbner basics are covered in standard textbooks: [AL, Ch.1–2], [BWK, Ch.5] or [CLO07, Ch.2–4]. A good reference in commutative algebra containing some advanced topics is [E95]. Refer to [E95, Ch.15] for elementary facts on flat deformation [BM, §1] and generic initial ideals [Gr, GrS].

The book [GPBLS] gives an elementary and beautiful account of commutative algebra on the computational and geometric aspects with lots of practical examples, and also provides coincidentally a guide to using the computer algebra system SINGULAR. The textbook [CLO05] illustrates fairly wide and recent applications, especially to geometry, combinatorics and coding theory. The book [S02] covers various substances relating to systems of polynomial equations and their applications even to statistics and economics. The monograph [S96] is about the interplay of computational commutative algebra and the theory of convex polytopes, which centers around the class of toric ideals. A concise textbook [Sch] could be regarded as a book of snapshots of commutative algebra via MACAULAY 2 [15], whose most part come from algebraic geometry with a backdrop of homological algebra. The book [EGSS] contains several materials using MACAULAY 2 from introductory chapters to mathematically difficult ones. A voluminous textbook [KR], using CoCoA[16] for actual computations, is an encyclopedia on computational commutative algebra which also contains so many tutorial sessions with interesting applications. Note that there is a database[17] containing data on various computer algebra systems, whose title is "Gröbner Bases Implementations: Functionality Check and Comparison".

The Shirshov's paper [Sh] on Lie algebras is also valuable to understand an original idea. The history and the main results of the Gröbner-Shirshov basis theory are reviewed in [BC, BK, BS]. We note that the notion of Gröbner-Shirshov pairs was introduced for the representations of associative algebras [KL] and an efficient algorithm for computing Gröbner-Shirshov pairs was constructed using a linear algebraic method [KLLP]. In addition, the computation of syzygies and the notion of Castelnuovo-Mumford regularity were generalized to representations of noncommutative algebras [KLPP].

[15]This is a computer algebra system, available at `http://www.math.uiuc.edu/Macaulay2` .

[16]This computer algebra system is available at `http://cocoa.dima.unige.it/` .

[17]Accessible at `http://www.ricam.oeaw.ac.at/Groebner-Bases-Implementations/` .

Account of Buchberger's algorithm in Korean

왜 문제가 되는지 감지하는 한가지 방법으로서 어떤 다변수 다항식을 여러 개의 다변수 다항식으로 나누는 상황을 생각해보자. 예를 들어 $xy^2 - x$을 $xy - y$로 나누고 $y^2 - x$으로 나누어보자. 이번엔 $xy^2 - x$을 $y^2 - x$으로 먼저 나눠보자. 이 경우 일변수 다항식의 나눗셈에서는 나타나지 않는 일이 발생한다. 즉, 나머지가 나눗셈 순서에 따라 달라지는 일이 생기는데, 우리가 기대하듯이 나머지가 다변수 다항식들로 생성되는 이데알 자체에만 의존하는 나눗셈을 얻기 위해서는 이데알의 "좋은" 생성원 집합을 구해야 한다.

여기서 최고차 이데알의 개념이 등장한다. 이것은 이데알의 모든 다항식의 최고차항으로 생성되는 단항 이데알이다. 임의의 이데알이 생성원으로 주어져 있을 때 생성원의 최고차항들로 만들어지는 이데알은 주어진 이데알의 최고차 이데알보다 일반적으로 작다. 그래서 주어진 이데알의 최고차 이데알이 무엇인지 아는 것이 필수적인데, 구체적으로 그 생성원을 구해야 한다. 그런데 자명한 문제는 주어진 이데알이 갖고 있는 다항식이 무한히 많다는 데 있다. 바꿔 말하면, 그 무수히 많은 다항식에 대해서 최고차항을 모두 고려하여 계산할 수가 없다. 최고차 이데알의 계산을 가능하게 해준 것이 부흐버거(Bruno Buchberger, 오스트리아 수학자, 1942–)의 1965년 박사학위 논문의 결과이다. 즉, *주어진 이데알의 무한히 많은 다항식을 고려하는 대신에 특정한 유한 개의 다항식에 대해서만 계산해주면 된다*는 사실을 증명해낸 것이다. 여기서 특정한 다항식이란 S-다항식을 말하는데 그것은 임의의 두 개의 생성원 각각에 적당한 단항식을 곱하고 서로 빼서 최고차항을 소거해준 새로운 다항식이다. 부흐버거는 유한 개의 S-다항식들을 고려하여 다항식의 나눗셈을 통해 나머지를 구했을 때 0 아닌 것들을 추가해주면 유한 과정 이후에는 모든 S-다항식들의 나머지가 0이 됨을 증명하였다. 이렇게 더 커진 새로운 생성원 집합의 최고차항들이 전체 최고차 이데알을 만들어낸다는 사실을 보인 것이다. 이같이 생성원의 최고차항들이 최고차 이데알을 만들어낼 때 우리는 그 생성원 집합을 *그뢰브너 기저(Gröbner bases)*라고 부른다.

부흐버거의 계산적 이론에 힘입어 우리는 다변수 다항식의 좋은 나눗셈을 할 수 있게 되었고, 이것은 선형대수에서 자유롭게 사용하고 있는 일차식 계산을 고차다항식의 경우로 획기적으로 발전시킨 것으로 볼 수 있다. 부흐버거의 방법은 좀 더 효율적인 알고리듬의 형태로 모든 계산 소프트웨어에 내장되어 있어서 복잡한 다항식 계산을 컴퓨터로 수행할 수 있게 되었다. 그뢰브너 기저 이론은 현대의 수학과 공학 뿐만 아니라 대수학적 계산이 요구되는 모든 분야의 핵심적인 바탕을 이루고 있다.

Bibliography

[이I] 이인석, *선형대수와 군: 학부대수학강의I*, 서울대출판부, 2005.

[이II] ———, *대수학: 학부대수학강의II*, 서울대출판부, 2008.

[Am] M. A. Armstrong, *Groups and Symmetry*, Undergraduate Texts in Math., Springer-Verlag, 1988.

[At] M. Artin, *Algebra*, Prentice Hall, 1991.

[Ch] A. Chambert-Loir, *A Field Guide to Algebra*, Undergraduate Texts in Math., Springer-Verlag, 2005.

[DF] D. S. Dummit, R. M. Foote, *Abstract Algebra*, 3rd ed., John Wiley & Sons, 2004.

[El] A. Elduque, *Introduction to Algebra*, Lecture Notes at Univ. Zaragoza, 2008.

[F] J. B. Fraleigh, *A First Course in Abstract Algebra*, 7th ed., Addison Wesley, 2003.

[G] J. A. Gallian, *Contemporary Abstract Algebra*, 2nd ed., D. C. Heath and Company, 1990.

[Hr] I. N. Herstein, *Topics in Algebra*, 2nd ed., John Wiley & Sons, 1975.

[Hm] J. F. Humphreys, *A Course in Group Theory*, Oxford University Press, 1996.

[L] S. Lang, *Undergraduate Algebra*, 3rd ed., Undergrad. Texts in Math., Springer-Verlag, 2005.

[Pi] C. C. Pinter, *A Book of Abstract Algebra*, MacGraw-Hill, 1982.

[R] J. Rotman, *Galois Theory*, 2nd ed., Universitext, Springer-Verlag, 1998.

- History of Mathemetics:

[BR] I. G. Bashmakova, A. N. Rudakov, *The evolution of Algebra 1800–1870*, Amer. Math. Monthly **102** (3) (1995), 266–270.

[Bl] E. T. Bell, *Men of Mathematics*, 1965. (안재구 옮김, *수학을 만든 사람들 上, 下*, 미래사, 1993.)

[D] J. Dorier, *A general outline of the genesis of vector space theory*, Historia Mathematica **22** (1995), 227–261.

[Ev] H. Eves, *Great Moments in Mathematics*. (허민, 오혜영 옮김, *수학의 위대한 순간들*, 경문사, 1994.)

[K] I. Kleiner, *A History of Abstract Algebra*, BirkHäuser, 2007. (김부윤, 정영우 옮김, *추상대수학의 역사*, 경문사, 2012.)

[V] V. S. Varadarajan, *Algebra in Ancient and Modern Times*, Mathematical World **12**, American Math. Soc. and Hindustan Book Agency, 1998. (김부윤, 송영무 옮김, *대수학의 어제와* 오늘, 경문사, 2005.)

- Linear Algebra, Number theory, and Topology:

[오이] 오정환, 이준복, *정수론*, 교우사, 1998.

[민] 이민섭, *정수론과 암호학*, 개정판, 교우사, 2008.

[CP] R. Crandall, C. Pomerance, *Prime Numbers: A Computational Perspective*, Springer-Verlag, 2001. pp. 225–349.

[Cu] M. L. Curtis, *Matrix Groups*, 2nd ed., Universitext, Springer-Verlag, 1984.

[Ka] D. W. Kahn, *Topology: An Introduction to the Point-Set and Algebraic Areas*, The Williams & Willkins Company, 1975.

[LF] R. Larson, D. C. Falvo, *Elementary Linear Algebra*, 6th ed., Brooks/Cole, 2010.

[NZM] I. Niven, H. S. Zuckerman, H. L. Montgomery, *An Introduction to the Theory of Numbers*, 5th ed., John Wiley & Sons, 1991.

- Advanced topics:

[김김] 김명환, 김홍종, 현대수학입문 - 힐베르트의 문제를 중심으로, 경문사, 2000.

[Bz] J. C. Baez, *The octonions*, Bull. Amer. Math. Soc. **39** (2002), 145–205.

[Ca] R. W. Carter, *Simple Groups of Lie type*, John Wiley & Sons, 1972.

[DL] W. Decker, C. Lossen, *Computing in Algebraic Geometry: A Quick Start using* SINGULAR, Algorithms Comput. Math. **16**, Springer-Verlag, 2006.

[Gd] M. Gardner, *Penrose Tiles and Trapdoor Ciphers*, W. H. Freeman, 1989.

[Gt] P. Garrett, *Abstract Algebra*, Lecture Note at Univ. Minnesota, 2007.

[H90] J. E. Humphreys, *Reflection Groups and Coxeter Groups*, Cambridge Studies in Advanced Math. **29**, Cambridge Univ. Press, Cambridge, 1990.

[Hg] T. W. Hungerford, *Algebra*, Graduate Texts in Math. **73**, Springer-Verlag, 1974.

[M] H. Matsumura, *Commutative Algebra*, 2nd ed., Benjamin/Cummings, 1980.

[S] R. Solomon, *A brief history of the classification of the finite simple groups*, Bull. Amer. Math. Soc. **38** (2001), 315–352.

[Wi] R. A. Wilson, *The Finite Simple Groups*, Grad. Texts in Math. **251**, Springer-Verlag, 2009.

- Applications:

[박] 박승안, 암호학과 부호이론, 2판, 경문사, 2008.

[유] 유석인, 이산수학, 개정판, 영지문화사, 1995.

[CDU] A. Cohen, J. Draisma, R. Ushirobira, *Group theory for Maths, Physics and Chemistry students*, Lecture Notes, 2002.

[MR] J. Meldrum, A. Ranicki, *Linear Algebra for Informatics*, `http://xmlearning.maths.ed.ac.uk/`, 2005.

[Pl] V. Pless, *Introduction to the Theory of Error-Correcting Codes*, 3rd ed., John Wiley & Sons, 1998.

[We] H. Weyl, *Symmetry*, Princeton University Press, 1952.

- Gröbner bases:

[AL] W. W. Adams, P. Loustaunau, *An Introduction to Gröbner Bases*, Grad. Studies in Math. **3**, Amer. Math. Soc., Providence, 1994.

[Ba] D. Bayer, *The Division Algorithm and the Hilbert Scheme*, Ph.D. thesis, Harvard University, 1982.

[BMo] D. Bayer, I. Morrison, *Standard bases and geometric invariant theory I. Initial ideals and state polytopes*, In: Computational Aspects of Commutative Algebra (L. Robbiano ed.), J. Symbolic Comput. **6** (1988), no. 2-3, 209–217.

[BM] D. Bayer, D. Mumford, *What can be computed in algebraic geometry?*, In: Computational Algebraic Geometry and Commutative Algebra (D. Eisenbud and L. Robbiano eds.), Proc. Conference (Cortona, 1991), Sympos. Math. **34**, Cambridge Univ. Press, Cambridge, 1993, pp. 1–48.

[BS87a] D. Bayer, M. Stillman, *A criterion for detecting m-regularity*, Invent. Math. **87** (1987), 1–11.

[BS87b] ———, *A theorem on refining division orders by the reverse lexicographic order*, Duke Math. J. **55** (1987), 321–328.

[BS88] ———, *On the complexity of computing syzygies*, In: Computational Aspects of Commutative Algebra (L. Robbiano ed.), J. Symbolic Comput. **6** (1988), no. 2-3, 135–147.

[BWK] T. Becker, V. Weispfenning, H. Kredel, *Gröbner Bases: A Computational Approach to Commutative Algebra*, Grad. Texts in Math. **141**, Springer-Verlag, New York, 1993.

[Be] G. M. Bergman, *The diamond lemma for ring theory*, Adv. Math. **29** (1978), 178–218.

[BJSST] T. Bogart, A. N. Jensen, D. Speyer, B. Sturmfels, R. R. Thomas, *Computing tropical varieties*, J. Symbolic Comput. **42** (2007), 54–73.

[BC] L. A. Bokut, Y. Chen, *Gröbner-Shirshov bases for Lie algebras: after A. I. Shirshov*, Southeast Asian Bull. Math. **31** (2007), 1057–1076.

[BK] L. A. Bokut, P. S. Kolesnikov, *Gröbner-Shirshov bases: from incipiency to the present*, Zap. Nauchn. Sem. POMI **272** (2000), 26–67 (in Russian). Translated in: J. Math. Sci. **116** (2003), 2894–2916.

[BS] L. A. Bokut, K. P. Shum, *Gröbner and Gröbner-Shirshov bases in algebra: an elementary approach*, Southeast Asian Bull. Math. **29** (2005), 227–252.

[B65] B. Buchberger, *An algorithm for finding the basis elements of the residue class ring of a zero dimensional polynomial ideal*, Ph.D. thesis, University of Innsbruck, 1965 (in German). Translated in: J. Symbolic Comput. **41** (2006), 475–511.

[B70] ———, *An algorithmic criterion for the solvability of a system of algebraic equations*, Aeq. Math. **4** (1970), 374–383 (in German). Translated in: [BW], pp. 535–545.

[B76a] ———, *A theoretical basis for the reduction of polynomials to canonical forms*, ACM SIGSAM Bull. **10** (1976), no. 3, 19–29.

[B76b] ———, *Some properties of Gröbner bases for polynomial ideals*, ACM SIGSAM Bull. **10** (1976), no. 4, 19–24.

[B79] ———, *A criterion for detecting unnecessary reductions in the construction of Gröbner bases*, In: Symbolic and Algebraic Computation (E. W. Ng ed.), Proc. EUROSAM (Marseille, 1979), Lecture Notes in Comput. Sci. **72**, Springer-Verlag, Berlin, 1979, pp. 3–21.

[B83] ———, *A note on the complexity of constructing Gröbner bases*, In: Computer Algebra (J. A. van Hulzen ed.), Proc. EUROCAL (London, 1983), Lecture Notes in Comput. Sci. **162**, Springer-Verlag, Berlin, 1983, pp. 137–145.

[B85] ———, *Gröbner bases: an algorithmic method in polynomial ideal theory*, In: Multidimensional Systems Theory: Progress, Directions and Open Problems in Multidimensional Systems (N. K. Bose ed.), Math. and its Applications **16**, D. Reidel Publishing Company, Dortrecht, 1985, pp. 184–232.

[B87] ———, *History and basic features of the critical-pair/completion procedure*, J. Symbolic Comput. **3** (1987), 3–38.

[B88] ———, *Applications of Gröbner bases in nonlinear computational geometry*, In: Mathematical Aspects of Scientific Software (J. R. Rice ed.), Proc. the IMA workshop (Minneapolis, 1987), IMA Vol. Math. Appl. **14**, Springer-Verlag, New York, 1988, pp. 59–87.

[B98] ———, *Introduction to Gröbner bases*, In: [BW], pp. 3–31.

[B01a] ———, *Gröbner bases: a short introduction for systems theorists*, In: Computer Aided Systems Theory (R. Moreno-Diaz, B. Buchberger and J. L. Freire eds.), Proc. EUROCAST (Las Palmas de Gran Canaria, 2001), Lecture Notes in Comput. Sci. **2178**, Springer-Verlag, Berlin, 2001, pp. 1–19.

[B01b] ———, *Gröbner bases and systems theory*, Multidimens. Systems Signal Process. **12** (2001), 223–251.

[BLRW] B. Buchberger, F. Lichtenberger, H. Rolletschek, F. Winkler, *An algorithm for constructing canonical bases of polynomial ideals*, ACM Trans. Math. Software **11** (1985), 66–78.

[BMö] B. Buchberger, H. M. Möller, *The construction of multivariate polynomials with preassigned zeros*, In: Computer Algebra (J. Calmet ed.), Proc. EUROCAM (Marseille, 1982), Lecture Notes in Comput. Sci. **144**, Springer, Berlin, 1982, pp. 24–31.

[BW] B. Buchberger, F. Winkler (Eds.), *Gröbner Bases and Applications*, Proc. Conference "33 years of Gröbner bases" (Linz, 1998), London Math. Soc. Lecture Note Ser. **251**, Cambridge Univ. Press, Cambridge, 1998.

[CKM] S. Collart, M. Kalkbrener, D. Mall, *Converting bases with the Gröbner walk*, J. Symbolic Comput. **24** (1997), 465–469.

[CLO05] D. Cox, J. Little, D. O'Shea, *Using Algebraic Geometry*, 2nd ed., Grad. Texts in Math. **185**, Springer-Verlag, New York, 2005.

[CLO07] ———, *Ideals, Varieties, and Algorithms: An Introduction to Computational Algebraic Geometry and Commutative Algebra*, 3rd ed., Undergrad. Texts in Math., Springer-Verlag, New York, 2007.

[DS] W. Decker, F.-O. Schreyer, *Computational algebraic geometry today*, In: Applications of Algebraic Geometry to Coding Theory, Physics, and Computation (C. Ciliberto, F. Hirzebruch, R. Miranda and M. Teicher eds.), Proc. NATO Advanced Research Workshop (Eilat, 2001), NATO Sci. Ser. II **36**, Kluwer Acad. Publ., Dordrecht, 2001, pp. 65–119.

[Dk] L. E. Dickson, *Finiteness of the odd perfect and primitive abundant numbers with n distance prime factors*, Amer. J. Math. **35** (1913), 413–422.

[E95] D. Eisenbud, *Commutative Algebra with a view toward Algebraic Geometry*, Grad. Texts in Math. **150**, Springer-Verlag, New York, 1995.

[EGSS] D. Eisenbud, D. R. Grayson, M. Stillman, B. Sturmfels (Eds.), *Computations in Algebraic Geometry with* MACAULAY 2, Algorithms Comput. Math. **8**, Springer-Verlag, Berlin, 2002.

[FGLM] J.-C. Faugère, P. Gianni, D. Lazard, T. Mora, *Efficient computation of zero-dimensional Gröbner bases by change of ordering*, J. Symbolic Comput. **16** (1993), 329–344.

[GM] R. Gebauer, H. M. Möller, *On an installation of Buchberger's algorithm*, In: Computational Aspects of Commutative Algebra (L. Robbiano ed.), J. Symbolic Comput. **6** (1988), no. 2-3, 275–286.

[GMNRT] A. Giovini, T. Mora, G. Niesi, L. Robbiano, C. Traverso, *"One sugar cube, please" or selection strategies in the Buchberger algorithm*, In: Proc. ISSAC (Bonn, 1991), ACM, New York, 1991, pp. 49–54.

[GMU] E. L. Green, T. Mora, V. Ufnarovski, *The non-commutative Gröbner freaks*, In: Symbolic Rewriting Techniques (M. Bronstein, J. Grabmeier and V. Weispfenning eds.), Proc. Workshop (Ascona, 1995), Progr. Comput. Sci. Appl. Logic **15**, Birkhäuser, Basel, 1998, pp. 93–104.

[Gr] M. L. Green, *Generic initial ideals*, In: Six Lectures on Commutative Algebra (J. Elias et al. Eds.), Proc. Summer school on commutative algebra (Bellaterra, 1996), Progr. in Math. **166**, Birkhäuser, Basel, 1998, pp. 119–186.

[GrS] M. L. Green, M. Stillman, *A tutorial on generic initial ideals*, In: [BW], pp. 90–108.

[Gl] G.-M. Greuel, *Computer algebra and algebraic geometry - achievements and perspectives*, J. Symb. Comput. **303** (2000), 253–290.

[GPBLS] G.-M. Greuel, G. Pfister, *A* SINGULAR *Introduction to Commutative Algebra*, 2nd ed., Springer-Verlag, Berlin, 2008 (with contributions by O. Bachmann, C. Lossen and H. Schönemann).

[GPS] G.-M. Greuel, G. Pfister, H. Schönemann, SINGULAR *3.0, A Computer Algebra System for Polynomial Computations*, Centre for Computer Algebra, University of Kaiserslautern (2005), `http://www.singular.uni-kl.de`.

[HW] C. J. Hillar, T. Windfeldt, *Algebraic characterization of uniquely vertex colorable graphs*, J. Combin. Theory Ser. B **98** (2008), 400–414.

[H72] J. E. Humphreys, *Introduction to Lie Algebras and Representation Theory*, Grad. Texts in Math. **9**, Springer-Verlag, New York, 1972.

[K-RW] A. Kandri-Rody, V. Weispfenning, *Non-commutative Gröbner bases in algebras of solvable type*, J. Symbolic Comput. **9** (1990), 1–26.

[KLLP] S.-J. Kang, D.-I. Lee, K.-H. Lee, H. Park, *Linear algebraic approach to Gröbner-Shirshov basis theory*, J. Algebra **313** (2007), 988–1004.

[KLPP] S.-J. Kang, D.-I. Lee, E. Park, H. Park, *A generalization of Castelnuovo-Mumford regularity for representations of noncommutative algebras*, J. Algebra **324** (2010), 631–651.

[KL] S.-J. Kang, K.-H. Lee, *Gröbner-Shirshov bases for representation theory*, J. Korean Math. Soc. **37** (2000), 55–72.

[KR] M. Kreuzer, L. Robbiano, *Computational Commutative Algebra 1 / 2*, Springer-Verlag, Berlin, 2000 / 2005.

[Le] V. Levandovskyy, *Non-commutative Computer Algebra for Polynomial Algebras: Gröbner bases, applications and implementation*, Ph.D. thesis, Universität Kaiserslautern, 2005.

[Li] H. Li, *Noncommutative Gröbner Bases and Filtered-Graded Transfer*, Lecture Notes in Math. **1795**, Springer-Verlag, Berlin, 2002.

[MaMe] E. W. Mayr, A. R. Meyer, *The complexity of the word problems for commutative semigroups and polynomial ideals*, Adv. Math. **46** (1982), 305–329.

[MM] H. M. Möller, F. Mora, *Upper and lower bounds for the degree of Gröbner bases*, In: EUROSAM 84 (J. Fitch ed.), Proc. ISSAC (Cambridge, 1984), Lecture Notes in Comput. Sci. **174**, Springer-Verlag, Berlin, 1984, pp. 172–183.

[Mo] T. Mora, *An introduction to commutative and noncommutative Gröbner bases*, Theoret. Comput. Sci. **134** (1994), 131–173.

[MoR] T. Mora, L. Robbiano, *The Gröbner fan of an ideal*, In: Computational Aspects of Commutative Algebra (L. Robbiano ed.), J. Symbolic Comput. **6** (1988), no. 2-3, 183–208.

[O] U. Oberst, *Multidimensional constant linear systems*, Acta Appl. Math. **20** (1990), 1–175.

[Sch] H. Schenck, *Computational Algebraic Geometry*, London Math. Soc. Sudent Texts **58**, Cambridge Univ. Press, Cambridge, 2003.

[Sh] A. I. Shirshov, *Some algorithmic problems for Lie algebras*, Sibirk. Math. Z. **3** (1962), 292–296 (in Russian). Translated in: ACM SIGSAM Bull. Commun. Comput. Algebra **33** (1999) no. 2, 3–6.

[Sta] R. Stanley, *Combinatorics and Commutative Algebra*, 2nd ed., Progr. in Math. **41**, Birkhäuser, Boston, 1996.

[S93] B. Sturmfels, *Algorithms in Invariant Theory*, Texts and Mono. in Symbolic Comput., Springer-Verlag, Wien, 1993.

[S96] ———, *Gröbner Bases and Convex Polytopes*, Univ. Lecture Ser. **8**, Amer. Math. Soc., Providence, 1996.

[S02] ———, *Solving Systems of Polynomial Equations*, CBMS Regional Conference Ser. in Math. **97**, Amer. Math. Soc., Providence, 2002.

[Tr] C. Traverso, *Hilbert functions and the Buchberger algorithm*, J. Symbolic Comput. **22** (1996), 355–376.

[U] V. Ufnarovski, *Introduction to noncommutative Gröbner bases theory*, In: [BW], pp. 259–280.

[Wa] U. Walther, *Algorithmic computation of local cohomology modules and the local cohomological dimension of algebraic varieties*, J. Pure Appl. Algebra **139** (1999), 303–321.

[W] F. Winkler, *On the complexity of the Gröbner-bases algorithm over* $K[x, y, z]$, In: EUROSAM 84 (J. Fitch ed.), Proc. ISSAC (Cambridge, 1984), Lecture Notes in Comput. Sci. **174**, Springer-Verlag, Berlin, 1984, pp. 184–194.

Index